Royal Botanic Gardens Kew

Souvenir Guide

Contents

Welcome to Kew

Our 250th anniversary year in 2009 has provided us with a wonderful opportunity to share Kew's remarkable heritage and the important contribution we are making to plant conservation in a time of rapid environmental change.

In 1759 Princess Augusta, mother of King George III, started an ambitious nine-acre garden around Kew Palace. Every generation has added to the charms and curiosities of Kew, now a major international visitor attraction comprising 132 hectares of landscaped gardens. But Kew today is much more, since it plays a pivotal role as a world leader in plant science and conservation.

As well as a site of World Heritage status, rich in history, Kew has the Earth's largest and most diverse botanical collections under its care. It is also an internationally respected centre of scientific excellence,

identifying and classifying plants; researching their structure, chemistry and genetics, collecting and conserving endangered species, maintaining reference collections and sharing this knowledge with others. Increasingly, we are reaching out across the world in partnerships with many organisations aiming to conserve plant life and chart a way forward that ensures a sustainable future for plants and people.

I hope you enjoy your time at Kew, and come again to see its richness and diversity through the seasons. Our 250th anniversary has led to new developments – there has never been a better time to visit. Above all, I hope you find inspiration to help care for the world's plants and fungi – the basis of all life.

Professor Stephen Hopper
Director, Royal Botanic Gardens, Kew

What to see at Kew

Glass Palaces

Kew contains many iconic buildings, not least its impressive collection of glasshouses. The Palm House and Temperate House are grand, Grade 1 listed, internationally-known symbols of Kew, while the modern Princess of Wales Conservatory and Davies Alpine House are award-winning designs, employing the latest technologies to control and regulate temperature and humidity levels.

See pages 6–31 for Kew's main glasshouses

Exotic Plantings

Delicate orchids and hungry carnivorous plants grow in the Princess of Wales Conservatory; Britain's oldest pot plant lives among tropical friends in the Palm House; a cycad, possibly the last of its species in the world, in the Temperate House; a thriving example of the 'pinosaur' Wollemi pine near the Orangery. Kew's beautiful and rare plants can be found throughout the gardens.

Royal Kew

From Kew Palace, built in 1631, the oldest building on site, to the magnificent Royal Crests on the wrought iron gates and the artful Royal 'follies' throughout the garden (including the famous Pagoda), the passions, interests and idioms of Britain's ancient Royal Family are everywhere.

See pages 32–41 for features of Royal Kew

Special Gardens

Treat all your senses in the Secluded Garden, enjoy the peace of the Japanese Landscape, see traditional medicinal plants in the seventeenth-century style Queen's Garden, or enjoy the fabulous collections of rhododendrons, magnolias, roses and bamboos. In all, fifteen unique gardens await you at Kew.

See pages 42–51 for details on Kew's special gardens

Trees

Kew grows a wonderful variety of trees from all over the world. The Arboretum stretches across two-thirds of the Gardens and contains over 14,000 specimens of more than 2,000 species and varieties, including enviable collections of oaks, ashes, cedars, and

redwoods. And at Kew you can do more than walk underneath and gaze up at the trees: by climbing the Xstrata Treetop Walkway you can experience the wonder of the canopy level, learn new facts about trees and enjoy fantastic views across the Gardens.

See pages 52–59 for more on Kew's trees

Water and Wildlife

Water features have figured strongly in Kew's history, and this is still the case today, with Kew's Lake, the Palm House Pond, the Waterlily Pond and views of the Thames. The Sackler Crossing is Kew's first permanent water crossing, enabling a closer appreciation of the lakeside planting and wildlife. Look out for waterfowl and overwintering wild birds by the lakeside in winter, great crested grebes in the spring and colourful kingfishers in the summer.

See pages 60–65 for more on water and wildlife

Art

Botanical art combines beauty with accuracy, revealing the natural world's often fragile beauty. There is no better place to view it than at the Shirley Sherwood Gallery, the world's first gallery solely dedicated to botanical art. Treasures from Kew's historic collection and from Shirley Sherwood's impressive private collection are displayed in exhibitions, changed three times a year. From autumn 2009, don't miss the newly-restored Marianne North Gallery. This gallery is filled with a unique collection of 832 colourful oil paintings by the Victorian amateur artist, depicting the many plants she encountered on her travels around the world.

See pages 66–71 for more on art at Kew

Seasonal Walks

From spring bulbs and a bluebell wood, to the sights and scents of summer in full bloom; from autumn colour, bark and berries, to winter snowdrops, wintersweet and witch hazels: the delights of Kew change with the seasons.

See pages 86–95 for seasonal walking routes

Glass Palaces

Nowhere else in the world is there such a collection of both historic and elegant glass buildings designed with such a high purpose – the presentation of magnificent plants for people to see, enjoy, investigate, and learn to respect.

Botanical science, horticulture, precision engineering and conservation meet together under the crystal domes and glittering towers of Kew's glass palaces.

Palms are second only to grasses in their importance to people. About 70% of all palm species are found in tropical rainforests, one of the most threatened habitats on Earth. The Palm House creates conditions similar to tropical rainforest; around a quarter of the palms planted here are threatened in the wild, as are more than half of the cycads, the 'living fossils' of the tropics.

The Palm House also contains many plants of great economic significance, grown for their yields of fruits, timber, spices, fibres, perfumes and medicines. Plants are grouped together in geographical areas, except in the centre, where the tallest need the extra height of the dome.

Palm House

AFRICA: South Wing

Kew is proud of its rare triangle palm (*Dypsis decaryi*) and bottle palm (*Hyophorbe lagenicaulis*) which is extremely rare in the wild. Other valuable plants originating in Africa include coffee bushes and the Madagascar periwinkle, from which anti-leukaemia drugs were developed.

South Africa is rich in cycads and one of Kew's specimens is the oldest pot-plant in the UK, brought here in 1775.

The World's Largest Seed
The fruit of Coco-de-mer (*Lodoicea maldivica*), illustrated above, takes seven years to develop and results in an unusual two-lobed seed, weighing up to 30 kg – the same as the average TV!

THE TALLEST PALMS: Centre Transept

Planting position under the great central dome of the Palm House is dictated by height, which cannot be reduced without killing the growing point, or 'heart' of the palm. The many-stemmed peach palm is often felled for its edible heart, 'coeur de palmier'. Also here are the babassu from central South America, the fast-growing queen palm and the very tall *Ravenea moorei* from the Comores Islands, possible the only example of the species in the UK.

THE AMERICAS

AFRICA

ASIA, AUSTRALASIA, THE PACIFIC

THE AMERICAS: Centre Transept

The Americas are the world's richest rainforest habitat. Caribbean palms and rare Mexican cycads have their own areas. Vitally important economic specimens including cocoa (shown left) are planted here. You can also see the Mexican yam, used to develop the contraceptive pill, and *Smilax utilis*, which is used to make the tonic drink sarsaparilla.

Parrot flowers (*Heliconia*), illustrated right, and the hybrid rose of Venezuela may display splashes of scarlet and in summer, the sweet scent of frangipani fills the air.

ASIA, AUSTRALASIA, THE PACIFIC: North Wing

This diverse collection includes shade-loving 'understorey' specimens, dwarf palms and climbing rattans. Familiar Asian fruit trees, such as the mango, are planted with the more

unusual breadfruit, jackfruit, and Indonesia's legendary durian. Sugar cane and spices such as ginger and pepper grow here.

Other notables include the spectacular jade vine and a wide selection of potted palms.

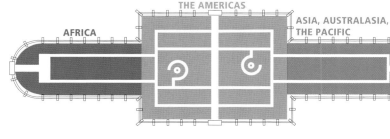

Planting and conservation

The Palm House represents one habitat – tropical rainforest. The plantings simulate its multilayered nature with canopy palms and other trees, climbers and epiphytes down to the shorter understorey plants and dwarf palms.

Many of these plants are endangered in the wild; some are even extinct. Their natural habitats are being cleared for agriculture, mining, or logging. Scientists at Kew are deeply involved in surveys and studies of rainforest species to assess habitat diversity and to develop methods for the sustainable growth and harvesting, as well as conservation.

All life depends on plants. If left unchecked, the current rate of destruction of the world's rainforests will see the total loss of this environment by the middle of this century, together with its peoples, its vast untapped potential for crops and medicines, and its vital moderating effect on the Earth's climate. Our planet will struggle to breathe.

Rainforest plants in the world economy

The Palm House contains many plants of economic importance. One of Kew's roles is research into the factors that make for sustainable cropping.

Look for:–

Rubber tree (*Hevea brasiliensis*)

Natural rubber has unique properties, making it important for many uses, from robust space shuttle tyres to delicate contraceptives. In 1876, Kew received 70,000 seeds collected from native Amazonian specimens. Only 2,800 germinated, but from them, the seedlings sent to Sri Lanka and Malaysia flourished, starting their massive rubber industries.

Rubber on tap
The white latex that flows in the inner bark of the rubber tree is tapped by a series of cuts (shown right).

African oil palm (*Elaeis guineensis*)

Native to tropical Africa, this, the most productive oil-producing tropical plant, is grown in plantations throughout the world's humid tropics. Palm oil pressed from the orange flesh makes soap, candles and edible products. Oil from the stone's kernel is used for edible oils and fats, soaps and detergents. Sap is tapped for palm wine, and the leaves and trunk make houses.

Cocoa (*Theobroma cacao*)

The Aztecs called it 'Food of the Gods', hence its scientific name deriving from 'theos' (gods) and 'bromos' (food). Originally a bitter beverage reserved for Aztec high society, the Spanish added sugar and vanilla to create its modern taste.

Giant bamboo (*Gigantochloa verticillata*)

Bamboos are known as 'friend of the people' in China and 'wood of the poor' in India. Giant bamboo (shown left) is used in house construction and paper-making and the young shoots are a table delicacy.

Coconut (*Cocos nucifera*)

Coconut probably originates from the islands of the western Pacific and eastern Indian oceans, spreading by ocean currents and human planting. Every part of the tree is used; the trunk for timber and the leaves for thatch, baskets and brooms. The nut provides food, drink and oil from inside and coir fibre and garden mulch from the outside.

Madagascar periwinkle (*Catharanthus roseus*)

Traditionally used in folk remedies for digestive complaints and diabetes, this pretty ornamental (shown right) is also important in the fight against cancer. Two of its six useful alkaloids can be used to treat leukaemia and Hodgkin's disease. Commonly cultivated, it is now rare in the wild.

Coffee (*Coffea*)

Coffee was first used as a paste of beans and oil, chewed for the caffeine effect. By the 15th century people were roasting beans to produce a drink. Arabica beans (*Coffea arabica*) make quality coffee, while the robusta (*Coffea canephora*) is used for instant coffee.

Pepper (*Piper nigrum*)

This woody climber (shown right), is native to India. Its fruits are possibly the world's most important spice. Black pepper comes from grinding the whole dried peppercorn; for white, the outer husk is removed. Pepper was first used as a cooking spice in India; coming to Europe in the Middle Ages to flavour and cure meat. It was also used with other spices to mask the taste of bad food.

Sugar cane (*Saccharum officinarum*)

Sugar cane originated in New Guinea, where ancient people used the stem for chewing. Today, raw cane sugar is a major world commodity and by-products such as molasses and bagasse (used for fuel) are valuable, too. Fuel alcohol produced from sugar is used in 50% of all cars in Brazil.

The Palm House – icon of Kew

Built in 1844–48 by Richard Turner to Decimus Burton's designs, the Palm House is Kew's most recognisable building, having gained iconic status as perhaps the world's most notable surviving Victorian glass and iron structure.

Borrowed technology

While the thinking was Burton's, the extraordinary engineering and construction work was Turner's. The technology came from shipbuilding and the design is essentially an upturned hull. The unprecedented use of light but strong wrought iron 'ship's beams' made the great open span possible.

The Grade 1 Listed Palm House was dismantled and restored in 1984–1989. It was re-opened by HM Queen Elizabeth the Queen Mother on 6 November 1990.

Marrying form and function

The Palm House was built for the exotic palms being introduced to Europe in Victorian times. The elegant design with its unobstructed space for the spreading crowns of tall palms is a perfect marriage of form and function.

Burton chose the location so that his building would be reflected in the Pond to the east. Pagoda Vista and Syon Vista radiate from this focal point of the Gardens.

The Oldest Pot-Plant in Britain

This cycad, *Encephalartos altensteinii*, was brought to Kew in 1775 and typifies the Palm House's original arrangement of plants displayed in pots. It needed special care when moved for the restoration in the 1980s.

The Campanile

The word 'campanile' comes from campana, the Italian word for bell, and refers to free-standing bell-towers. But Kew's elegant Campanile, designed by Decimus Burton and erected in 1847, was actually built to disguise a chimney.

Originally, the Palm House boilers were in the basement, piping hot water under iron gratings to heat the plants above. Smoke was led away to the Campanile through a 150 m tunnel, which also featured a rail track, enabling coal to be hauled to the boilers. The system was not a success, and in 1867 discreet chimneys were installed in the Palm House wings, although the boilers remained in the basement until the 1950s.

The Marine Display

The Marine Display in the Palm House basement emphasises the importance of marine plants and, through displays in 19 tanks, recreates four major marine habitats: coral reefs, estuaries and salt marshes, mangrove swamps and rocky shorelines.

Relatives large and small

Algae, which include all seaweeds, provide half of the world's oxygen supplies and absorb vast amounts of carbon dioxide.

Algae vary enormously in shape and size. The tiny, single-celled phytoplankton, include the world's smallest plant, *Chlorella*, which is only 0.003 mm in diameter. Among the macro-algae is giant kelp (*Macrocystis*) which, at up to 100 metres in length,

is one of the world's largest plants. Blue-green algae (*cyanobacteria*) are believed to be the world's oldest plants.

Active alae

Algae are a vital first link in the ocean's food chain. They are also largely unexploited by man, though Japanese nori and Welsh laverbread are well-established foods. Seaweed has been long used as land fertiliser and animal fodder in coastal communities. Today, algal products include a variety of gels, such as agar, alginates and carrageenan; used in ice cream, growing media for microbial culture, paints and very effective wound dressings.

Green scum and red tides

Micro-algae as single cells are usually invisible to the naked eye, but occur in vast numbers as green scum on ponds and red tides on beaches and in reservoirs. Under a microscope, they show a huge variety of beautiful shapes.

Depth of colour

Seaweeds are divided into groups by colour – green, brown and red. Green algae live in shallow waters up to 10 metres deep. Brown algae can survive in water up to 30 metres deep, while red algae can survive in water to a depth of 100 metres. All algae need light for photosynthesis, and each group contains different coloured pigments to absorb the available daylight at different depths.

The Queen's Beasts

Ten stylised heraldic figures, sculpted of Portland Stone, stand in front of the Palm House; replicas of those which stood at Westminster Abbey during the 1953 coronation of HM Queen Elizabeth II. Created by the late James Woodford, they were presented to the Gardens in 1956. The beasts are selected from the armorial bearings of the Queen's forebears to illustrate her royal lineage. They are shown as they appear when facing the Palm House with **The Lion of England** on the far left.

The Falcon of the Plantagenets

The Red Dragon of Wales

The Black Bull of Clarence

The Unicorn of Scotland

the spectacle of the magnificent building. A fountain was added in 1853, and the current sculpture of Hercules wrestling the river-god Achelous in 1963. This bronze cast was made by Crozatier for George IV in 1826, and used to stand on the East Terrace of Windsor Castle. It is based on an original sculpture by Bosio in 1814, in the Jardin de Tuilieres, Paris.

Two white marble Chinese Lions can be found 'guarding' the Victoria Gate side of the Pond. They are thought to date from the Ming Dynasty of AD1368–1644 and were presented to Kew by Sir John Ramsden in 1958.

The Palm House Pond

The Palm House Pond was part of the major transformation of Kew in 1845, under William Nesfield and Decimus Burton. Once a small remnant of Kew's original lake, it was enlarged and reshaped to provide a water feature in its own right, and a balance to the formal parterres. It also reflected the entire length of the Palm House in its waters adding to

The White Lion of Mortimer

The White Horse of Hanover

The Yale of Beaufort

The White Greyhound of Richmond

The Griffin of Edward III

The Palm House Parterres

The massed flower beds that present themselves in sections immediately to the front and side of the Palm House are a reflection of the formal garden parterres traditionally set within the grounds of a Victorian stately home. Less intense now than the 17th century French and German designs on which they were first patterned, they nevertheless display a remarkable range of colour and focused seasonal planting.

The original layout of the parterres was drawn up by painter-turned-landscaper William Nesfield in 1845 and included what is now the Rose Garden. They were designed to set off the grandeur of the Palm House and became the subject of some disagreement between the garden 'artist' and the Director Sir William Hooker, who was a pragmatic botanist and simply wanted the plants in species order. Fortunately for Kew a compromise was reached.

Seasonal planting themes are designed and implemented by expert horticultural staff. The arrangements have become less formal over the years, though nevermore so than during the first world war when the beds were dug up and replanted with onions!

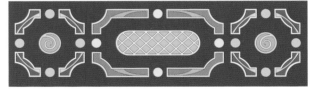

Diagram depicting the original layout of the parterres.

Above, left and right: the summer bedding displays outside the Palm House are impressive whether viewed in their entirety or up close.

Waterlily House

The Waterlily House (next to the Palm House) is another classic Kew building, again with ironwork by Richard Turner. Built in 1852, it was then the widest single span glasshouse in the world, designed specifically to house the huge attraction of the age, the giant Amazonian water lily *Victoria amazonica*.

No one could believe this giant tropical sister of the little 'frog pads' in a garden pond could grow leaves two metres across and have enough buoyancy to support a baby! Named in honour of the queen, it was considered the central living attraction at Kew throughout Victoria's reign.

The great water lily responded poorly to its new home and in 1866 the house was converted to an Economic Plant House. It was converted back to its original use in 1991, and thanks to modern technology is now the warmest and most humid environment at Kew.

In summer, the *Nymphaea* water lilies and a giant *Victoria cruziana* put on a beautiful display, while lotus and papyrus thrive in the hot, humid conditions. The corner beds contain economically useful plants such as rice and lemon grass. High up, there are spectacular gourds such as loofahs, hedgehog gourds and wax gourds.

The world's largest surviving Victorian glass structure, the Temperate House is another of Decimus Burton's amazing designs and another of Kew's 39 unique listed buildings. At 4,880 square metres it is the largest glasshouse at Kew, twice the size of the Palm House.

Tender woody plants from the world's temperate regions have always been a major part of the collection at Kew. In Victorian times the pace of collecting meant that the Orangery and many other houses quickly became overcrowded so, in 1859, it was decided to build another glasshouse to complement the Palm House.

The planting here is in geographical zones as intended in Burton's original design, though the scheme now represents many more regions than there were originally.

Temperate House

The world's largest indoor plant is the Chilean wine-palm (*Jubaea chilensis*) in the centre of the Temperate House, which is 16 metres high – and still growing! It was grown from seed and there is a replacement nearby, ready for the time when this huge wine palm no longer fits into the roof space.

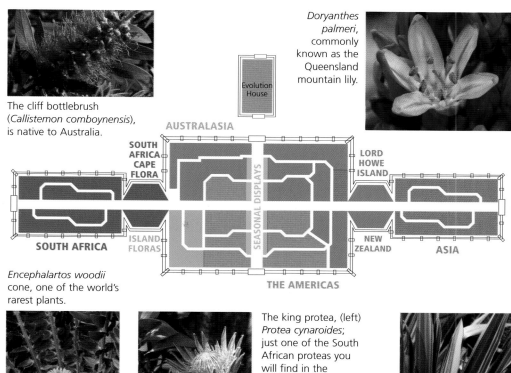

The cliff bottlebrush (*Callistemon comboynensis*), is native to Australia.

Doryanthes palmeri, commonly known as the Queensland mountain lily.

AUSTRALASIA

SOUTH AFRICA CAPE FLORA

Evolution House

LORD HOWE ISLAND

SEASONAL DISPLAYS

SOUTH AFRICA

ISLAND FLORAS

NEW ZEALAND

ASIA

THE AMERICAS

Encephalartos woodii cone, one of the world's rarest plants.

The king protea, (left) *Protea cynaroides*; just one of the South African proteas you will find in the Temperate House.

Encephalartos woodii

Phormium, (right) a New Zealand native, was one of the first plants to be discovered by Captain Cook.

Temperate zones

The Temperate House collection includes spectacular specimens that are deservedly admired, but it represents much more than that. Among the plants on display are endangered island species being propagated for reintroduction to their native lands, such as *Hibiscus liliiflorus* from Rodrigues Island and *Trochetiopsis erythroxylon* from St. Helena. There are also many plants of economic importance such as the date palm, tea, quinine and chillies.

The Temperate House holds an extensive collection of temperate American plants, including fuchsias and salvias. The central section holds the Australian collection, with grass trees, the delightful 'kangaroo's paws' and an array of banksias, named after their collector, Sir Joseph Banks.

Life-saving tonic

From the bark of the cinchona tree (shown left) comes quinine, used as a flavouring in Indian tonic water and taken worldwide as a life-saving drug to counter malaria.

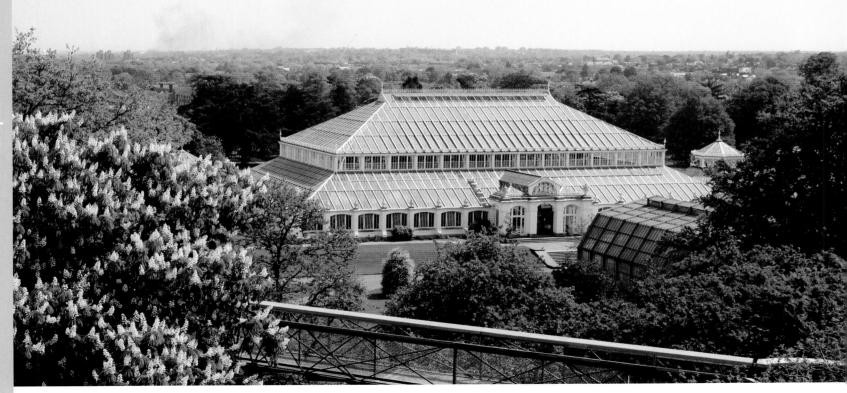

Bird of paradise

The comman name of *Strelitzia reginae* (shown right) derives from the bird-like appearance of the flowers. You can see five *Strelitzia* species of varying size and form on display in the Temperate House.

And who's the rarest of them all?

A cycad, *Encephalartos woodii* (shown right) was presented to Kew by the Natal National Park. This is the rarest plant in the Gardens, and is extinct in the wild. All cycads bear male (pollen) and female (seed) cones on different plants. The Kew tree is a lone male and it remains an extraordinary challenge to encourage it to cone (shown on p.19).

An ambitious plan

By the mid-19th century, the need for a large temperate greenhouse had become overwhelming. In 1859, the Government allocated £10,000 to build the Temperate House and directed Decimus Burton to prepare designs.

The plants needed good ventilation, so the house was designed in straight lines. The glazing bars were of wood, not iron, for easy repair and to aid heating, with a decorated cornice, a very Victorian flourish, at the eaves.

Building on a promise

Work began in 1860. The octagons were completed in 1861, the centre section in 1862 and foundations for the wings were part laid when work was stopped in 1863. The Treasury called a halt when the construction bill came to £29,000 and work did not resumed until August 1895. The south wing was finished in 1897, then the contractor went bankrupt. The north wing was completed by another in 1898.

Moving moments

When the octagons were ready, the contents of the 'New Zealand House' were transferred. Australian plants, 'unhappy' trees from the Orangery and palms from the overburdened Palm House were all established in the new house.

The central interior had 20 oblong beds, with lines of araucarias, palms, and other tall plants in the middle, side beds of rhododendrons, acacias and magnolias, and smaller plants in pots and boxes on benches.

The public were first admitted in May 1863, although the building was far from finished.

The chosen site in the centre of the Arboretum was raised two metres by creating a huge terrace of sand and gravel excavated from, and creating, the Lake.

The Temperate House provides a dramatic backdrop to Kew's immensely popular Summer Swing music festival.

The Evolution House is a fascinating journey through eons of plant development. The first life forms, 3.5 billion years ago, were *cyanobacteria* – living mats on the surface of structures called *stromatolites*.

Land plants emerged in the Silurian period. *Cooksonia* was one of the first vascular plants – which means a plant with conducting channels or 'veins' for carrying sap.

The Evolution House picks three major periods – the Silurian, Carboniferous and Cretaceous – and includes a coal swamp showing the giant club mosses and horsetails from an estimated 300 million years ago. Cycads appeared about 100 million years later followed by the conifers and flowering plants.

Among species featured here are the world's largest horsetail (*Equisetum giganteum*) and the high-climbing fern *Lygodium*, whose fronds can grow to over 30 m long.

Giant woodlouse (*Arthropleura*) may have lived at the time of the coal swamps. This model (shown left) can be seen in Evolution House.

Coal – The 'stone' which burns

In the Carboniferous period clubmosses, tree ferns and giant horsetails flourished in warm damp conditions in extensive areas of wetlands. When these plants died, they fell into an airless morass where, instead of rotting away, they eventually formed the hardest of the so called 'fossil fuels': coal.

Coal is formed when plant debris is buried under high temperature and pressure. Earthquakes or volcanoes submerge the plants, various chemicals are driven out, the structure becomes fossilised and eventually turns to coal. The process probably started 360 million years ago and went on through stages, forming first peat and then lignite, or soft brown coal, before hard, black coal.

Evolution House

The most complex of Kew's public glasshouses commemorates Princess Augusta who married Frederick, Prince of Wales in 1736 and who founded the Gardens. It was opened by Diana, Princess of Wales on 28 July 1987.

Ten different environments cover the whole range of conditions in the tropics, ranging from the dry scorching heat of the desert to moist tropical rainforest – under one roof. Water features strongly in the humid zones, with pools of impressively large fish and the famous giant Amazonian water lily.

Plants of great economic importance such as pepper, bananas and pineapples grow here, as well as orchids and carnivorous plants.

Princess of Wales Conservatory

Massive attraction

The giant Amazonian waterlily (*Victoria amazonica*), shown left, was a huge attraction in Victorian times. Close relatives *Victoria cruziana* and *Victoria* 'Longwood hybrid' are just as popular today, and you can see the latter here from April to October.

Sophronitis cernua, on display in the orchid section.

The Dry Tropics Zone is home to many cacti (shown above) and succulants. A desert backdrop is created by a Sherman Hoyt painted diorama.

Ferns (shown left) come from both tropical and temperate regions, so they appear in two conservatory zones.

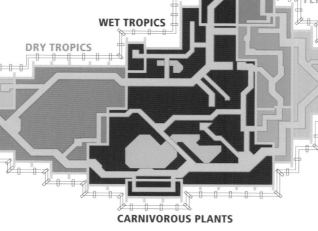

ORCHIDS

FERNS

SPECIAL DISPLAYS

WET TROPICS

DRY TROPICS

CARNIVOROUS PLANTS

Titan arum

With its foul stench and rare flowerings, the Titan arum (*Amorphophallus titanum*), shown left, never fails to make an impression.

Slippery slope

Insects are attracted to the wide open mouth of the pitcher plant by glistening droplets of nectar. But as soon as they climb inside to reach it, they lose their footing on the slippery walls and plop into a pool of digestive juices at the bottom. But some insects have learned to survive. Their larvae are immune to, and swim in, the strong fluid, and live on scraps the plant does not eat.

Lounging lizards

Some lizards, including *Acanthosaurus armata*, are used as biological control in the Princess of Wales Conservatory, where they thrive on a diet of insect pests. If you are lucky – and very quiet – you might catch a glimpse of one.

Two main climate zones, the Dry Tropics and Wet Tropics, occupy most of this conservatory. There are eight more different micro-climates, each created for the needs of a particular plant group. All plants are shown as naturally as possible, with ferns clinging to dripping rock faces, and climbers on columns. Paths on different levels bring visitors close to the plants so they can appreciate the subtle details of the vegetation.

The Dry Tropics Zone represents arid regions from around the world. Here, there are plants that have many different ways of dealing with the lack of water, conserving it with waxy skins or fleecy jackets, or storing it in succulent stems.

Although coming from opposite sides of the world, the agaves and aloes show many similarities in adaptations for survival.

Cacti flowers are typically spectacular and complex, as the vibrant flowers on this *Mammillaria* specimen show.

There are also collections of the unique native plants from the Canary Islands and Madagascar, many of which are threatened as their habitats in the wild are destroyed.

The Seasonally Dry Zone is an enclosure watered sparingly in the winter containing plants from the East African deserts and savannah, all adapted in different ways. The baobab stores water in its thick trunk, while acacias shed leaves in the dry season to conserve moisture.

Nearly 20% of the world's people live on the edge of desert. Drought and inappropriate farming techniques bring danger to semi-arid ecosystems. Kew scientists are learning more from this living collection, studying the use of plants as sustainable crops for food, fuel, fodder and medicine, and to help slow the rate of desert encroachment.

A cunning camouflage

Plants called 'living stones' or 'mimicry plants' (*Lithops*) are so well camouflaged in their surroundings that they seem safe from animal grazing – until they flower.

The Wet Tropics Zone is maintained with the high humidity typical of rainforests. The planting also reflects the lighting conditions; low at floor level where species such as the marantas, with their patterned leaves, are flourishing. Also on show are wild forms of many familiar house plants, such as the Swiss cheese plant, African violet and begonias.

Personalised climates

There are a further eight different micro-climates, each created for the needs of a particular plant group.

Even higher humidity and also cool shade is required for plants from the high cloud forests, so a separate enclosed area is specially set aside for them.

Many carnivorous plants are found in cool, well-lit areas and, since they often grow on poor soils, need to supplement their diets by catching insects. The Venus flytrap and pitcher plant are getting hungry in the east of the conservatory.

Orchids have two distinct zones to provide the growing conditions that different types require. A hot, steamy zone features tropical epiphytic or air-rooting varieties with spectacular showy flowers and specific adaptations to an aerial environment in the rainforest canopy. The cooler orchid zone suits species with their roots in the earth of tropical mountainous regions.

Ferns, too, come from both tropical and temperate regions and there are two separate areas for ferns which reflect their different needs.

An award-winning building that works on many levels

The challenge of building a major new glasshouse at Kew was both technical and aesthetic. It had to replace no fewer than 26 elderly glasshouses with one sophisticated building. Aesthetically, the site was exceptionally sensitive, being close to both the Palm House and Orangery.

The keynote for the design was the highest possible energy efficiency allied to the lowest possible maintenance. The high humidity and temperatures needed to support tropical plants mean a glasshouse like this is under attack all the time. A durable design was needed.

With its stepped and angled glass construction, without sidewalls and with most of its space below ground, the conservatory is a most effective collector of solar energy. The volume is relatively low in relation to its floor area so that temperatures within the individual zones may be altered quickly.

Above all, the Princess of Wales Conservatory looks stunning. The old adage that 'form follows function' is as perfectly demonstrated today as it was over 160 years ago with Burton's Palm House. It looks in place in its surroundings, even in such imposing company. It is beautifully landscaped, too, blending into the Rock and Woodland Gardens to the east and south and surrounded by the mature trees of the Arboretum to the north and west.

Ten different environments, ranging from the extreme temperature range of the desert to the moist heat of mangrove swamps, are controlled by computer to provide a flexible mix of heat, humidity and light. Sensors on walls and in the beds report the exact environmental conditions to the computer which commands heat to flow, roof louvers to open, or mists to spray to increase humidity. It seems as if the building itself is alive to its charges.

The boiler room is underground. Also beneath the conservatory are two 227,000 litre (50,000 gallon) storage tanks for rainwater collected from the roof slopes and used, after filtration and ultraviolet treatment, for irrigation.

Among its many awards the Princess of Wales Conservatory was presented with the Institution of Structural Engineers special award in 1987 and the highly prestigious Europa Nostra Award for Conservation in 1989.

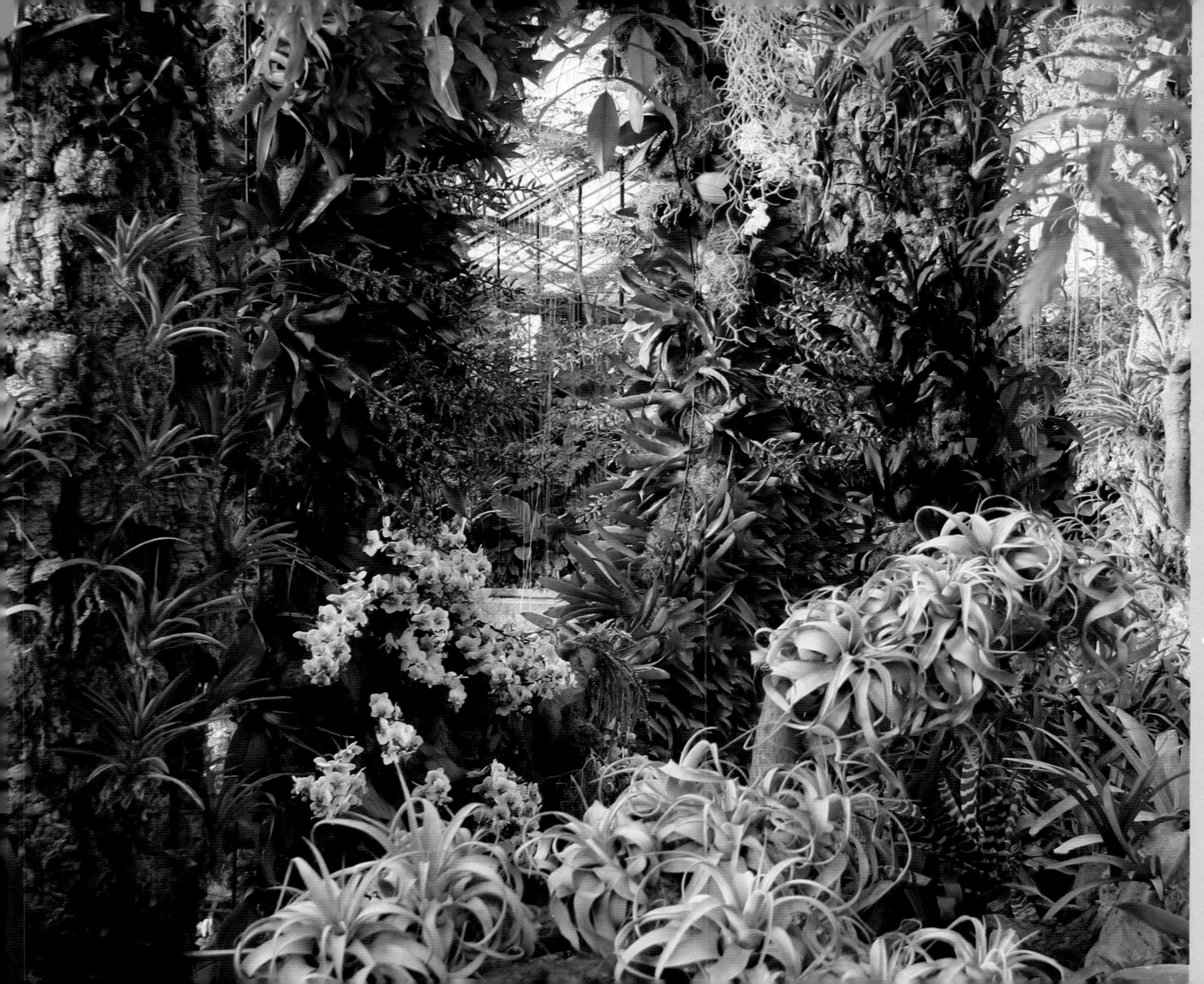

Opened in early 2006, this elegant new house is designed to provide the right conditions for alpine plants, keeping them dry in the winter and cool in the summer. The distinctive shape of the Alpine House allows cool air to be drawn in at the base as warm air escapes through vents in the roof. Shading helps to keep the air cool, while the 12 mm laminated glass with its low iron content allows 90% of light through. The plants require free draining soil and thrive in the landscape of Sussex sandstone recycled from the former Alpine House. The Davies Alpine House was awarded a prestigious RIBA award for architecture in 2006.

Davies Alpine House

Cool cooling

The cooling system uses a basement chamber as a 'heat sink'. During warm days, air is drawn into the system from outside and passed through a concrete labyrinth which absorbs the heat before the air enters the glasshouse. The objective is to 'flood' the plants with cooled air. At night, when ambient temperatures are cooler, the air flow continues into the labyrinth, while any warmth in the basement exhausts to the outside. In this way the labyrinth is cooled ready to supply cool air to the glasshouse the following day. The house stays within temperature range of 0–28°C.

Iris cycloglossa

A shade more or less

A sail type shading system follows the line of the glasshouse. Visually pleasing, it is best thought of as four huge Japanese fans, hinged at the middle of each wall. A mechanised system of ropes, pulleys and winches automatically draws up the fans when shade is needed. Each side of the glasshouse shading is independent so that only the 'sunny side' is shaded.

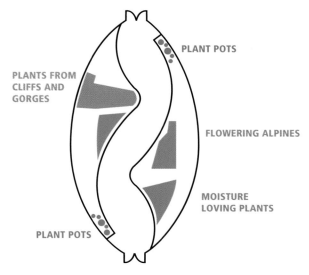

PLANT POTS

PLANTS FROM CLIFFS AND GORGES

FLOWERING ALPINES

MOISTURE LOVING PLANTS

PLANT POTS

The Rock Garden

The rock garden was designed in 1882 to mimic a Pyrenean mountain valley. It has now been redesigned to include a central bog and cascade. The Rock Garden represents six global regions, and visitors can walk through the garden by the waterfalls to enjoy the extraordinary variety of plants on display.

The alpines have well-drained soil and a grit mulch to prevent water splashing on to their leaves. Mediterranean plants are in the sunniest spots, while the woodland plants enjoy the shade and damp conditions created by water features.

Tulipa sylvestris subsp. *australis*

Dry and wild

From wild open mountainsides to Arctic shorelines, alpines grow in a huge variety of places. True alpines have short wet summers and long dry winters. They flower and set seed in a small window of opportunity between these two extremes – often as soon as water becomes available – and are highly adapted to their extreme environment.

All change

The superb plants in the Alpine House, such as *Lewisia rediviva* (shown right) are just a fraction of the collection. The display is regularly changed as plants come into flower.

Royal Kew

There are many Royal residences in Great Britain, some highly prestigious, some richly historic and some warmly appealing. But to find all three superlatives together not in a castle, nor mansion but in a graceful garden, is truly unique.

So tread where the Kings and Queens of Great Britain have trod, see where they took their leisure, where they cultivated powerful friendships – and plants – discussed the fate of nations, drew plans to confound their enemies, and sometimes even their friends.

Kew Palace

Built in 1631 for a rich merchant Samuel Fortrey (whose initials are on the front decoration) this is the oldest building on the Kew site. Originally known as the Dutch House, it was leased by the Royal Family in 1728 and became home to the three elder daughters of King George II. In 1731, George II's son Frederick, Prince of Wales, leased a house opposite known as Kew Park, and converted it into a large Palladian palace, the 'White House'.

George III grew up in the White House and lived there until 1802, when he purchased Kew Palace and demolished the White House. It was at Kew Palace that his wife Charlotte gave birth to George IV, and here that they found retreat during the King's well known, but hushed up, bouts of 'madness'.

After Queen Charlotte died in 1818, Kew Palace was closed. In 1896, Queen Victoria agreed to Kew's acquisition of the Palace, providing there was no alteration to the room in which Queen Charlotte died. In 1898, the Palace passed to the Department of Works and opened to the public for the first time. Today the Palace is in the trust of Historic Royal Palaces.

Kew Palace has recently been restored to its 1804 condition. Its interiors provide a contrast to typical expectations of royal palaces and it provides a fascinating insight into Georgian taste and style. Previously closed areas of the Palace are open, including rooms on the second floor that have remained untouched and unseen since the Royal Family's departure nearly 200 years ago.

For more details go to *www.hrp.org.uk*

The Ice House

The Ice House, in use by 1763, is an interesting reminder of the days before mechanical cooling. A thick wooden door opened to the north-facing entrance tunnel, and the whole brick-lined structure was covered with earth for insulation. Ice was collected during winter months and stored packed in straw, for use in drinks and iced desserts throughout the year. Collecting ice in the winter was a miserable, unpopular job and extra beer was given to estate staff in compensation.

Nosegay Garden

This sunken garden is lined on three sides by a pleached walk of laburnum (shown below). The plants were selected primarily for their reputed medicinal qualities. Seventeenth-century herbals by authors such as John Gerard, John Tradescant and John Parkinson were used for research, and quotations from these adorn the plant labels revealing, for instance, that common marigold comforts hearts and spirits, and borage relieves sorrow and increases 'the joye of the minde'.

Queen's Garden

Planting and conservation

Secreted behind Kew Palace is a charming 17th century-style garden. The word 'style' is used advisedly because, despite its historic appearance, it was only conceived in 1959 by Sir George Taylor, then Director of the Royal Botanic Gardens, and officially opened by HM Queen Elizabeth II ten years later.

A key element is the parterre enclosed in box hedges and planted with lavender, cotton lavender, purple sage, variegated iris, lady's mantle and Russian sage. Standing in the pond to the rear of this a copy of Verocchio's 'Boy with a Dolphin', the original of which is in Florence. Other attractive features are the sunken Nosegay

Garden, full of sweet-smelling and useful flowers, and a traditional vegetable garden.

The plants in the Queen's Garden are exclusively those grown in Britain before and during the 17th century. Their labelling differs from Kew's norm, since they include not only today's botanical name and family, but also the common name in the 17th century, a virtue, or quotation from a herbal (plant book) and the author's name and date of publication.

There is a great detail to see here – a wrought iron pillar from Hampton Court Palace, a chamomile chair, intricately plaited laburnums forming an arch, pleached hornbeams making a hedge on tall stems, a mound covered in clipped box, with a gazebo and a fine smoke tree.

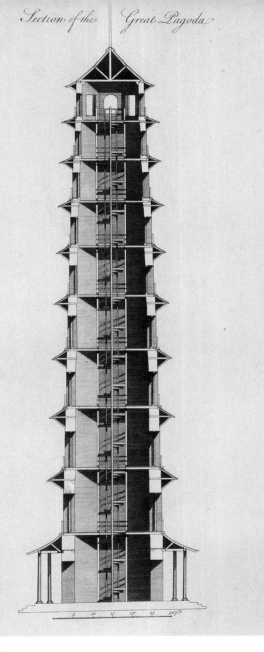

Section of the Great Pagoda

The Pagoda

One of Kew's famous features, the Pagoda is one of 25 ornamental buildings designed by Sir William Chambers for Kew when it was a Royal estate .

The Pagoda was completed in 1762 as a surprise for Princess Augusta. It was one of several Chinese buildings made for the Kew estate by Chambers to add a final flourish to the end of the garden and it was originally flanked by a Moorish Alhambra and a Turkish Mosque.

The ten-storey octagonal structure, constructed by local builder Solomon Brown, is 50 m high and was, at that time, the most accurate reconstruction of a Chinese building in Europe (although purists argue that pagodas should always have an odd number of floors). It tapers, with successive floors from the first to the topmost being 30 cm less in both diameter and height than the preceding one.

The original building was intended to be very colourful: Chambers, who had lived in China, wanted the roofs to be made of green and white varnished iron plates, the banisters to be a mix of blue, red and green and 80 gilded dragons to decorate the roof corners with a gold finial to top it all off. However, iron plates were replaced by slate and sadly, the dragons vanished 22 years later in 1784, during maintenance.

The Pagoda has undergone no fewer than 24 colour scheme changes since then. In 1843, Decimus Burton wanted to restore the Pagoda to its former glory, but the £4,350 cost was considered too high. It has been opened to the public only a handful of times in its history, most recently in 2006 for Kew's heritage year festival.

Bombs away!

Contemporaries of Chambers wondered if such a tall building would remain standing, though it had been "built of very hard bricks". Its sturdy construction was proved in World War II when it survived a close call from a stick of German bombs exploding nearby. This was ironic, since at the time, holes had been made in each of its floors so that British bomb designers could drop models of their latest inventions from top to bottom to study their behaviour in flight.

Although normally closed, the Pagoda was opened to the public during 2006. Spectacular views (shown left) awaited those who managed the 253-step climb to the top.

Sir William Chambers (1723–1796)

Born to Scottish parents in Gothenburg, William Chambers went to live in China as an employee of the Swedish East India Company, which sparked his interest in Chinese architecture. At the age of 26, he studied to become an architect and trained classically in Rome. In 1757 he became Princess Augusta's official architect and architectural tutor to her son, the future George III. He supervised the buildings at Augusta's five establishments and, over a five year period, designed 25 decorative buildings for Kew. Many, such as the Mosque, the Alhambra, the Palladian Bridge and Menagerie have disappeared, but the Orangery, the Ruined Arch, the Temple of Bellona and, above all, the Pagoda, are permanent reminders of his talents. He was knighted in 1771 by King Gustav III of Sweden for a series of drawings of Kew Gardens, and George III allowed him to use the title in England.

The Orangery

This 1761 building is the earliest at Kew designed by Sir William Chambers and also the largest classical style building in the Gardens. Princess Augusta's coat of arms was placed over the central bay of the facade in the 1840s, along with the Royal Arms and escutcheons with the monogram 'A' in honour of Queen Adelaide, wife of William IV. Designed to hold orange trees, the light levels inside the building were too low to grow plants, even after glass doors were added in 1842. Today it is an elegant café-restaurant.

The Nash Conservatory

Of a classical stone 'Greek temple' design, this is the oldest of Kew's 19th century glasshouses. It was originally one of four pavilions designed by John Nash for the gardens at Buckingham Palace, but William IV moved this one to Kew in 1836. It was then adapted by Sir Jeffry Wyatville, the designer of King William's Temple.

Having served many uses over the years, the Nash Conservatory is now fully restored to take its place among Kew's iconic buildings. It is used for a number of art exhibitions, and contains the World heritage Site commemorative plaque.

Museum No. 1

In 1841 Sir William Hooker set about persuading the administrators of the need for a museum of economic botany. Decimus Burton adapted part of an existing building into a museum, which opened to instant acclaim in 1847. But within a few years it became clear the museum was too small. Burton designed a new museum – Museum No. 1 – which opened in 1857.

Museum No. 1 hosts the Plants+People Exhibition, which shows highlights from Kew's Economic Botany Collection. Its focus is on sustainable and medicinal uses of plants, and on conservation.

Queen Charlotte's Cottage is maintained and administered separately from Kew by Historic Royal Palaces and opening times are limited. See www.hrp.org.uk for more details.

Queen Charlotte's Cottage

The origins of this cottage may well have been the single storey building provided for the Menagerie Keeper. There is no doubt that Queen Charlotte was given the building in 1761 when she married George III and that she extended the property upwards by a floor and also in length.

This picturesque house in 'cottage ornée' style was used by the family as a shelter whilst walking, and for snacks and occasional meals. The large ground floor room had Hogarth prints on the walls, removed in the 1890s but replaced in 1978. A curved staircase leads to the picnic room, with painted flowers climbing the walls and bamboo motif pelmets and door frames. These charming murals were painted by Princess Elizabeth herself, which emphasises the family's feeling for the cottage.

The cottage remained private until 1898, when Queen Victoria ceded it and its 15 hectares to Kew to commemorate her Diamond Jubilee. The grounds had rarely been visited; trees were lying where they had fallen, but one condition the Queen made was that the area should be kept in its naturalistic state. This was supported by the Linnaean Society on behalf of ornithologists to maintain the area as a suburban haven for birds. That is how today's Conservation Area first came into being and how one of London's finest bluebell woods is kept intact.

Gates, Statuary and Sculpture

Four of Kew's gates are Grade II listed: the Main Gate, the Lion Gate and Lodge, the Unicorn Gate, and the Victoria Gate. The Main Gate, designed by Decimus Burton in 1845 and completed the next year, signified a changed of attitude in Kew's management: where previously visitors were required to be escorted through gardens by staff, now they could enter independently. The other three listed gates were all built in the second half of the nineteenth century.

The Lion and Unicorn statues standing above their respective gates remind visitors that in addition to the Royal Beasts paraded before the Palm House, commissioned sculpture appears at almost every vantage point in the gardens.

Of particular royal, not to mention scientific, interest is the sundial with marble pedestal (shown left) moved from Hampton Court and

placed on Kew Palace lawn by William IV in 1832. He wished to commemorate ground breaking astronomical observations made there 100 years before by James Bradley, Professor of Astronomy at Oxford, which led to the discovery that the earth 'wobbled' on its axis (nutation).

By modern contrast a similar commemorative scientific sculpture – a double helix in steel (Bootstrapping DNA, 2003, Charles Jencks) stands in front of the Jodrell laboratory – marking the 50th anniversary of the discovery of DNA.

The Chambers collection

The Ruined Arch, Temple of Aeolus, Temple of Bellona and Temple of Arethusa were all designed and constructed by Sir William Chambers. They are all Grade II listed buildings.

The Ruined Arch (1759) is a charming mock ruin – these were very fashionable at that time.

The Temple of Aeolus (1760-63), shown right, Aeolus was the mythical king of storms and the four winds, inventor of sails and a great astronomer. The temple was originally built of wood and once had a revolving seat to provide a panoramic view of the whole Kew estate. It was rebuilt in stone by Decimus Burton in 1845.

The Temple of Bellona (1760), shown above, is named after the Roman goddess of war. Behind its Doric facade is a room decorated with plaques bearing the names of British and Hanoverian regiments which distinguished themselves in the Seven Years' War (1756–63).

The Temple of Arethusa (1758) is named after Arethusa, a nymph attendant of Diana the Huntress. Today this temple is the setting for Kew's war memorials and is adorned with wreaths every November 11th.

Kew's Royal Follies

'Follies', ornamental architectural features of little practical use, were a fashion in grand garden design in the 17th and 18th century. Made to set off or finish a perceived 'romantic landscape' they were often contrived ruins or classical temples. The idea was to prompt a reflective or historic romantic feeling as the visitor viewed the garden.

Kew's follies are an eclectic collection and should be looked at in the context of their period. Past plans show that many have come and gone and thanks should be given for those that remain as treasured features of this World Heritage Site.

When they were built, these buildings represented all the fervent interest and excitement of knowledge gained, at least by the aristocracy, in a widening world. China and the Islamic world were opening up. There was a revival of classicism through the 'Grand Tour' taken by 18th century gentry. This passion for the newly-discovered influenced the design and development of the Gardens.

King William's Temple

Built in 1837 by Sir Jeffry Wyatville to complement Chambers' Temple of Victory (no longer standing), this stone building (below) with its Tuscan porticos contains iron plaques commemorating British military victories.

Decimus Burton (1800–1881)

Burton, the son of a London builder, was a brilliant architect from an early age. At Kew, he designed the Palm House in conjunction with Richard Turner, the Temperate House, Museum of Economic Botany, and Main Gate, worked with William Nesfield on the design for the Broad Walk and rebuilt the Temple of Aeolus.

Kew's Special Gardens

To wander in someone else's garden, well kept and full of surprising and exotic plants, is a rare pleasure. For a start you usually have to know the gardener to get invited! But here at Kew you are more than welcome to explore and enjoy not one but fourteen unique and different gardens within the whole compass of Kew.

Each one is themed and planted to offer you a different impression, a different experience and a special way to relax in green company.

Duke's Garden

The walled Duke's Garden (named after the 2nd Duke of Cambridge, whose land this once was) is planted partly for fun and partly for science. Familiar shrubs and herbaceous perennials around the lawns draw interest throughout the year. But this tranquil area is also home to the Exotic Border, a place for experiments in plant hardiness. Cannas and gingers are among the tender plants left out for the winter, and their resistance to the cold will determine how the border evolves. The Gravel Garden, sponsored by Thames Water, shows a range of attractive plants needing less water than traditional English garden choices.

Outside the walls, the Duchess Border carries the Lavender Species Collection, with well-known garden lavenders and a number of tender species rarely grown in the UK. Various Mediterranean specimens are trialled here also to determine their hardiness in the south of England.

Grass Garden

Designed in 1982, the Grass Garden contains 550 species and rising! Grasses (*Gramineae*, also known as *Poaceae*) are some of the most economically important plants, supplying food directly as cereals and indirectly as cattle fodder. They are the basis of many alcoholic drinks and are also much used in building – straw thatch and bamboo – while sorghum and sugar cane (grown in the Waterlily House and Palm House) can be used to produce petrol substitutes.

Visit in early summer for temperate annual grasses and cereals, and autumn for the perennials and their seed heads. Other display beds have British native grasses, and tropical and temperate cereals. The layout of the Grass Garden has recently been redesigned, with new paths to allow visitors to get closer to the plants. Of all the ornamentals, *Miscanthus sinensis* is the most notable, for both its past and its future potential. Grown from a single batch of seed, it is now a widely used architectural plant, and is being investigated by Kew scientists as an alternative fuel source.

Aquatic Garden

The Aquatic Garden was opened in 1909, inspired by the sunken gardens at Hampton Court Palace. The original design even included hot water pipes to give the plants an early seasonal start! This practice died out long ago, and today all the plants are grown at ambient outside water temperatures.

The garden contains some 40 varieties of water lilies in the central tank, with 70 other water plant species such as sedges and rushes in the corner tanks. The flowering rush (*Butomus umbellatus*) and the bog bean (*Menyanthes trifoliata*) are fine specimens.

This Garden is best visited in high summer, when the waterlilies and aquatic plants are in full flower.

The Bonsai House

At the northern end of the Order Beds – furthest from the Victoria Plaza – there is a small glasshouse with a wonderful collection of bonsai trees. 'Bonsai', is a growing technique that produces miniaturised specimens which still retain the individual characteristics of the full-sized species.

Secluded Garden

This intimate cottage style garden appeals to all the senses and features poems on sight, scent, hearing and touch. Behind earth mounds, pleached limes form a hedge to circle a spiral fountain (pleaching is trimming so that hedge tops grow out above bare stems). A stream here is bordered with waterside plantings and scented flowers.

Even poor weather days are interesting when rain splashes off the giant leaves of *Gunnera tinctoria* and wind rustles mysteriously through a tunnel of whispering bamboos. The planting here closely resembles a private garden, with apple, pear and quince trees, roses, hostas, lilies, irises and salvias.

Rose Pergola

The Rose Pergola was built in 1959 as part of a range of ground improvments to mark Kew Gardens' bicentennial. It runs along the central path of the Order Beds, not far from the old pergola it replaced.

The Pergola is covered with a variety of cultivated climbing roses selected for their profusion and length of flowering. Some of the glorious roses grown here cannot be grown in the main rose garden. The surrounding walls provide shelter for many plants, including *Actinidia kolomikta* with its striking pink, red and white spring foliage.

Students' Gardens

The students' vegetable gardens by the Order Beds are a definite attraction for those vistors who are competitive gardeners!

At Kew's School of Horticulture, established in 1963, students can take the three-year Kew Diploma of Horticulture. The Diploma course offers training in amenity and botanical horticulture. The aim is to provide an opportunity to study scientific and technical subjects at degree level, plus gain vital practical experience. Students are employees of the Kew and receive payment throughout. There is also a programme of international courses in herbarium techniques and botanic garden management, and research degrees can also be taken here.

First year diploma students are allocated vegetable plots (by the Order Beds) to maintain for one season. The plots are assessed monthly on their range of crops, crop protection, tidiness and overall crop health. Many of the crops are compulsory, but students can choose two additional vegetables as well as two cut-flowers or companion plants. The flowers attract insect pollinators to improve the setting of seed of the peas and beans.

What's in a name?

Plants have both common names and a scientific Latin name. The same plant may be called by many different common names, in one country or many. Here in the UK, for example, bird's-foot trefoil (*Lotus corniculatus*) is also known as hen and chickens, Tom thumb, granny's toenails, cuckoo's stockings, and Dutchman's clogs. But any botanist, from London to Tokyo, knows what is meant by *Lotus corniculatus*.

The scientific name can be recognised everywhere. The advantage of a Latin scientific name is that as Latin is a dead language it cannot change its meaning, but, at the same time, is still highly descriptive. And as there are no political or nationalistic overtones with Latin, it is accepted throughout the world.

How you can recognise the plants

This typical label shows the system used in the Royal Botanic Gardens. Other plant labels may contain additional symbols or information.

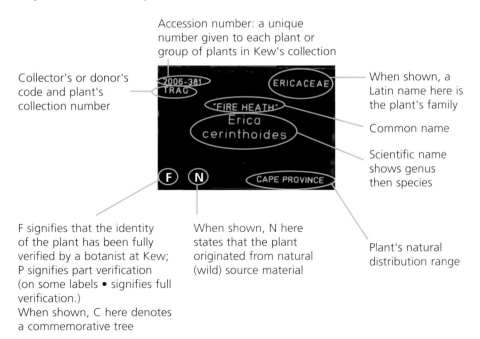

Accession number: a unique number given to each plant or group of plants in Kew's collection

Collector's or donor's code and plant's collection number

When shown, a Latin name here is the plant's family

Common name

Scientific name shows genus then species

F signifies that the identity of the plant has been fully verified by a botanist at Kew; P signifies part verification (on some labels • signifies full verification.)
When shown, C here denotes a commemorative tree

When shown, N here states that the plant originated from natural (wild) source material

Plant's natural distribution range

Order Beds

The 104 Order Beds (also known
collectively as the 'Systematic Garden')
were originally introduced by Sir Joseph Hooker
in the late 1860s and are a 'living textbook' of over 3,000
flowering plants. These are almost exclusively temperate
herbaceous dicotyledonous plants (plants with two seed leaves).
They are arranged systematically in family groups, so they can be easily located
and noted for study and comparison. This process, the science of understanding the
relationship between plants, is known as taxonomy. Some traditional family groupings have
recently been revised as a consequence of extensive research done at Kew, and elsewhere, into
the genetic make-up of plants. These changes are reflected in the latest arrangement of the Order
Beds. Science apart, the Order Beds are a stunning display in full summer bloom.

Woodland Garden

Here, a charming garden exactly replicates nature's design, being in three layers. High up, a deciduous tree canopy of oaks and birches shades the middle layer of deciduous shrubs such as maples and rhododendrons which, in their turn, protect the ground-cover layer, including hellebores, primulas, Himalayan blue poppies and North American trilliums.

Spring is the most colourful season here, when a range of woodland bulbs such as anemones, dog's tooth violets and scillas are in flower. In summer the trees provide leafy cover for the shade-loving perennials.

Winter Garden

The Winter Garden is laid out for interest on grey days. Specimens are set off against an evergreen backdrop, perfect for those which flower on bare wood. Here, *Mahonia* x *media* 'Winter Sun', and the chocolate-scented *Azara microphylla* surround wintersweets, viburnums, flowering quince, cornelian cherries, witch hazel and willows with their yellow catkins. There are bulbs, too, planted under the woody specimens, with winter aconite, snowdrops and windflowers left to naturalise. Conservationists will note *Abeliophyllum distichum* which is endangered in its native Korea.

Lilac Collection

Many say the lilac is the most elegant and colourful of all early summer flowering shrubs. Although historical literature and poetry indicates that Kew had lilac gardens in the past, this is newly developed. It was completely renovated in 1993 and replanted in 1997, though it still contains many of the original hybrids created at Kew from the 1900s. The Collection has 105 specimens of hardy lilacs ranging from the oldest species to modern cultivars and hybrids, and it is laid out over 10 separate beds according to cultivation and breeding history.

Mediterranean Garden

To the west of King William's Temple, leading out from a central green with an ornamental sundial, is a special collection of Mediterranean habitat plants, shrubs and herbs. The area is designed to recapture the unique feel of this part of the world both visually and aromatically. You can walk among the scented plants on the minor paths through the garden. Many love the collection of rock roses (*Cistus*) and the impressively large, and familiar, Yuccas.

The Rose Garden

The Rose Garden was created in 1923 and is a major draw for visitors, being hard by the Palm House – and in glorious bloom from June to August. Visiting the Rose Garden will offer many ideas to the garden enthusiast. Cluster-flowered and large-flowered roses are arranged by colour; red nearer the Palm House, contrasting against the white building; shading out to lighter colours at the perimeter, where white and yellow roses are set against the green hedges and vistas. The semicircular holly hedge and sunken areas are historically important, being the remains of an 1845 design by William Nesfield.

Each of the 54 beds contains a different cultivar of rose, all available through British rose growers. Ten of the beds illustrate the hybridisation of roses through the centuries. The roses displayed consist of floribunda, hybrid Ts, shrub and Old English roses which are all arranged by the colour of their flowers.

A major project is in progress in the Rose Garden, to take the beds and 'well areas' back to the original Nesfield design. Kew staff are working with David Austin Roses to devise a planting plan featuring pergolas and climbing frames. When completed the garden will encircle three sides of the Palm House and include statuary, walkways, and a 'history of the rose' area: a truly remarkable setting for an already remarkable building.

The Japanese Gateway and Landscape

Chokushi-Mon (The Gateway of the Imperial Messenger) is a four-fifths size replica of the Karamon of Nishi Hongan-ji in Kyoto, built for the Japan-British Exhibition in London in 1910. It is built in the architectural style of the Momoyama period in the late 16th century. Fine carvings depict flowers and animals, with intricate panels portraying a Chinese legend about a wise master and his devoted pupil.

The Japanese Landscape lies in three areas. The main entrance leads into a 'Garden of Peace', calming and tranquil, with stone baths and a gently dripping water basin. The slope to the south is a 'Garden of Activity' symbolising the natural grandeur of waterfalls, hills and the sea. The third area, the 'Garden of Harmony', links the other two and represents the mountainous regions of Japan, using stones and rocky outcrops interplanted with a wide variety of Japanese plants.

The Bamboo Garden

The Bamboo Garden makes the most of bamboos' variety of forms, stem colours and leaf shapes. Bamboos are actually grasses – ranging from giant poles, through wispy variegated species, to fountains of leaves from the pendulous varieties. In the centre of the Bamboo Garden there is a traditional Japanese Minka House.

The Japanese Minka

Until the middle of the 20th century, many Japanese country people lived in wooden houses called minka. They had sturdy wooden earthquake-resistant frames, mud-plastered walls and thatched roofs. They are uniquely 'green' buildings as everything used in their construction comes from plants or directly from nature.

Kew's minka was originally built around 1900 in Okazaki City. The house was donated to Kew as part of the Japan 2001 Festival.

Magnolia Collection

The first exotic magnolia was introduced to this country in the 17th century. Many of these deciduous and evergreen shrubs and trees with their spectacular pink, cream or white flowers do very well here, as can be seen from April to June in a planting near the rhododendrons and azaleas.

Rhododendron Dell

Rhododendrons are one of the largest and showiest of flowering shrubs, with great variation in size, habit and form. Over 700 specimens are planted in the Dell, with some unique hybrids found only here. This is not a natural valley, but a creation of the famous 'Capability' Brown, who carved what he called the 'Hollow Walk' in 1773.

Azalea Garden

Anyone looking for azaleas to grow for their blaze of late spring colour will be spoilt for choice at Kew. Twelve different groups of azalea hybrids have appeared since the very first Ghent hybrids in the 1820s and the Garden's two circles of beds arrange the plants in date order, leading up to today's varieties from eastern America and Holland. All the species planted in the Azalea Walk leading into the Garden belong to a group of deciduous azaleas from North America and Japan.

In the 1850s, Sir Joseph Hooker started the collection with specimens gathered on his Himalayan expeditions. The oldest is *Rhododendron campanulatum*, the most highly-scented are *Rhododendron* x *kewense*, named after the Gardens, and *Rhododendron* x *loderi* 'King George'. *Rhododendron* 'Cunningham's White' puts on a brilliant show each spring thriving in the local conditions, shaded by trees and kept humid by the nearby Thames.

Trees at Kew

There are over fourteen thousand trees planted at Kew, forming woodland glades, species groves, conservation collections and eye lines for grand vistas. Some are as old as the Gardens themselves and a great many can not be found anywhere else in the British Isles.

View them in their homeland habitats in the glasshouses or from high up among the boughs and branches on the Treetop Walkway, or stroll along a shaded avenue under their magnificent spreading splendour. In so many ways Kew's trees are the great, silent guardians of our heritage.

Kew's Rhizotron and Xstrata Treetop Walkway

Kew's Rhizotron and the Xstrata Treetop Walkway takes you under the ground and then up in the air, bringing you close to trees in a way that will take your breath away.

Walking in the air

Walking 18 m above the ground through a tree canopy of sweet chestnuts, limes and deciduous oaks, discover the birds, insects, lichens and fungi that rely on trees and look down, as they do, on Kew's 132 hectares and the London skyline. Woodland is the natural wild habitat of much of London and today still covers 4.5% of Greater London.

But if it does take your breath away – remember that is what trees do, in reverse, to the planet. An acre of trees removes about 2.6 tonnes of carbon dioxide a year, returning oxygen to the air. Forests cover 30% of the planet, but 7.3 million hectares (0.2%) of forest is lost per year.

Trees are cool, too: just a 10% increase in urban greenery can cut the temperature in a city by up to four degrees centigrade in summer.

Strolling along this 200 m walkway is a lofty, tranquil and surprisingly intimate experience. Designed by Marks Barfield Architects, who designed the London Eye, the pioneering structure of the Xstrata Treetop Walkway is ingeniously based on the 'Fibonacci ratio' – a fundamental creative sequence in nature observed by this 12th century Italian mathematician. The walkway structure is interconnected and supported in a robust but eye-pleasing natural manner as a consequence.

London underground

Kew's Rhizotron (taken from the Greek rhiza, meaning root) offers an equally unique opportunity, delving into the underground world of trees. Entering through an apparent crack in the ground, you can explore the lively natural world beneath the trees, including the relationships between tree roots and micro-organisms in the soil. Mechanically and electronically animated, the Rhizotron artistically helps you to get to the root of things.

Tree Mate

Tree Mate is a mobile phone based game. You may download (free of charge) photos and facts onto your mobile phone via bluetooth. Then once on the walkway you can explore five favourite trees, and then submit your comments to the Tree Notes section of the website. You can find Tree Mate at the end of the Rhizotron, before you climb the stairs to the walkway.

Fallen oak

A fallen tree demonstrates the amazing inner mechanism of such massive plants. Near the Xstrata Treetop Walkway you can explore the inner-workings of trees by peering into an open section of a huge fallen oak tree. Large sculptures of microscopic elements of the tree surround the oak, offering a more detailed view of root hair fibres and leaf pores.

As you walk around the treetops look out for the following tree favourites:

Common beech *Fagus sylvatica*

Beech trees can grow up to 40 m in height, over as long as 300 years. In tightly packed woods, the tree will grow straight with few side branches in a bid to reach the light. Leaves grow in an overlapping arrangement, which, while efficient for the tree, shades the ground and can prevent rain from reaching it. So if the floor of the wood you're walking through comprises little more than fungi and rotting leaves, it is probably a beech wood.

Horse chestnut *Aesculus hippocastanum*

Commonly known as a conker tree, the horse chestnut was traditionally planted as an ornament. With its spectacular spring and early summer blossom and autumn conker crop, it is a tree for all seasons. Full-grown trees provide shade for livestock in the summer, but, conkers aside, it serves surprisingly little practical purpose as its wood is soft, weak and does not burn well.

Turkey oak *Quercus cerris*

Native to south east Europe, with an excellent reputation for timber, the Turkey oak was introduced to Britain in 1735. At the time, English oak (*Quercus robur*), was widely used for all sorts of construction, from houses to ships, and it was hoped the Turkey oak would be a good alternative. It grew at a prodigious rate to an impressive 40 m high but the wood proved to be too brittle for widespread use. Despite this, Turkey oak's adaptability and unusual mossy acorn cups have made it a popular addition to the many oaks now found growing in Britain.

Northern red oak *Quercus rubra*

Red by name and red by nature, this oak is native to North America and is famous for its spectacular display of ruddy autumn leaves. As well as looking good, the red oak produces excellent timber and its wood is valued highly in its native country. Despite being thought of as a typically English tree, there are actually more than 450 species of oak – only two of which are native to Britain. The red oak is one of many from the Americas.

Sweet chestnut *Castanea sativa*

The sweet chestnut is a deciduous tree, widespread across Europe. Originally native to south-eastern Europe and Asia Minor, fully-grown it is typically an impressive 20–35 m in height with a trunk about 2 m in diameter. Its durable wood is used to make furniture, barrels, fencing and roof beams. In autumn it produces edible nuts that have been eaten for centuries. Attractive catkins hold the flowers of both sexes.

English oak *Quercus robur*

Unrivalled king of the forest, the English oak is synonymous with strength, size and longevity, and its name suggests its robust and sturdy nature. Oaks can grow up to 40 m and live in excess of 1,000 years. Ever since iron tools were first made, this mighty tree has been cut down for its strong and durable timber, even though it can take 150 years for an oak to be ready for use in construction. Until the middle of the 19th century, when iron became the material of choice for building ships, thousands of oaks were felled every year. It was estimated that it took 2,000 trees to make a single ship, and eventually laws were passed to protect it.

The Arboretum

Exotic trees have been collected at Kew since its earliest days, but it was during the stewardship of Sir Joseph Hooker (the son of Kew's first Director, Sir William Hooker) that today's Arboretum took shape.

Woodland Glade

The Woodland Glade is sheltered by mature oaks, underplanted with shrubs for summer and late autumn colour. There's a sense of calm by the charming Waterlily Pond with its profusion of aquatic life as well as beautiful plants.

Berberis Dell

Created between 1869 and 1875, this was once a gravel pit and is Kew's third-biggest excavation after the Lake and Rhododendron Dell. It has a secluded character and a large collection of *Berberis* and *Mahonia*. A wonderful Emily Young sculpture makes a striking focal point.

Cherry Walk

Stretching from the Rose Garden to the Pagoda, Cherry Walk is a superb collection of Japanese ornamental cherry trees. Replanting of the original 1935 walk was completed in 1996 and now, 22 trees in 11 matched pairs of different cultivars make a spectacular show in spring.

Holly Walk

Holly Walk is an important historical feature, originally laid out in 1874 by Sir Joseph Hooker; most of the hollies are now over 135 years old. Holly Walk is the largest, most comprehensive collection of mature hollies in cultivation. The picturesque berries vary from red to black or white and are at their best in November and December.

The Pinetum

Kew's first Pinetum was planted nearly 250 years ago, near the Orangery, and the second was created near the Palm House. Today's Pinetum, at the southern end of the Gardens, was started in 1870 by Sir Joseph Hooker.

Highly coloured

Autumn colour reaches its peak with the vibrant leaves of the maples between the Temperate House and Victoria Plaza, while still more colour comes from the berries and foliage on various trees of the Rosaceae (rose) family; whitebeams, rowans, hawthorns and crab apples.

Kew's heritage trees

The Arboretum contains specimen trees dating from as far back as the early 18th century to newly planted rarities from recent Kew collecting expeditions. See the map at the front of the guide for the locations of these trees.

Sweet chestnut, *Castanea sativa*
Early 18th century

This ancient tree is probably the oldest in the Gardens. The several sweet chestnuts here are thought to be the remnants of the woodland planting that lined Love Lane and separated the Kew and Richmond estates.

The Lucombe oak, *Quercus* x *hispanica* 'Lucombeana' 1765

The first Lucombe Oak was created in 1762 in a Mr Lucombe's Exeter nursery, as a cross between Turkey oak and cork oak (*Q. cerris* x *Q. suber*). Kew's specimen was planted around 1773, but was moved 20 m in 1846 to make way for the new avenue of holm oak (*Q. ilex*) along Syon Vista.

Turner's oak, *Quercus* x *turneri* 1798

This semi-deciduous oak was bred in a Mr Turner's Essex nursery in the late 18th century, as a cross between English oak and holm oak (*Q. robur* x *Q. ilex*). During the Great Storm in October 1987, the whole root plate of the tree lifted and loosened. But instead of damaging the tree this action rejuvenated it, prompting the ongoing decompaction programme for the Arboretum's mature trees.

'The Old Lions'

'The Old Lions' are some of the few remaining trees known to have been planted in the original garden of Princess Augusta around 1762. They are the maidenhair tree (*Ginkgo biloba*), the Japanese pagoda tree (*Styphnolobium japonicum*); the Oriental plane (*Platanus orientalis*); the false acacia (*Robinia pseudoacacia*); and the *Zelkova carpinifolia* in the Herbarium paddock.

Corsican pine, *Pinus nigra* subsp. *laricio* 1814

This large Corsican pine may be Kew's unluckiest tree. Early in the 20th century, a light aircraft crashed into it, taking out the top of the tree. Since then it has been struck by lightning at least twice, the last time in 1992, leaving only scars on the trunk as a reminder.

Tulip trees: *Liriodendron tulipifera* 1770s, *Liriodendron chinense* 2001

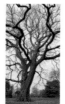

The North American species of tulip tree, *Liriodendron tulipifera*, was introduced to Britain in 1688. A specimen planted in the 1770s still flourishes in the Azalea Garden. The Chinese tulip tree, *Liriodendron chinense*, a far superior species, was introduced by Ernest Wilson in 1901, but is now very rare. Saplings grown from seeds collected during a recent expedition to China were planted in 2001 to recreate the Arboretum's old 'Tulip Tree Avenue'.

Chestnut-leaved oak, *Quercus castaneifolia* 1846

The chestnut-leaved oak was introduced to Britain from the Caucasus and Iran as seed in 1843. The specimen behind the Waterlily House was Kew's first, planted in 1846. At over 30 m in both height and spread, it is unrivalled as Britain's biggest and finest chestnut-leaved oak.

Indian horse chestnut, *Aesculus indica* 'Sidney Pearce' 1935

This fine tree, named after Sidney Pearce, the Arboretum's Assistant Curator in 1935, is a particularly good flowering form of *A. indica*. The mature specimen on the Little Broad Walk was transplanted from a collection near the Pavilion restaurant along with two *A. indica* specimens. Another three trees have since been added to create a colourful avenue from the Main Gate.

Stone pine, *Pinus pinea* 1846

Stone pines have been cultivated in Britain and continental Europe for some 2,000 years for pine nuts – the seeds from their cones. This specimen, planted in 1846, developed its unusual shape as a result of being kept in a pot for many years. In its pot, the stone pine became bonsaied, but once planted out it grew to form the shape it has today. Another name for the stone pine is 'umbrella' pine due to its parasol appearance when grown on a single trunk.

Water and Wildlife

If you fear that an historic London garden, carefully landscaped with exotic plants from around the world is too restrictive – that things may be too well-controlled to let local nature thrive – take a walk on the wild side of Kew, and be surprised.

Down by the lakeside, riverside and in the woodland there is wildlife in abundance. Here nature has a free hand. And Kew makes sure it stays that way, deliberately. Though in one or two places we may have helped our wild wonders show off just a little, so you can see and enjoy them.

The Lakeside and Sackler Crossing

Kew's present lake, occupying two hectares in the Arboretum, was commissioned by the then Director, Sir William Hooker, to provide an 'open flow of water through a portion of the pleasure grounds'. Work started in 1856 and the lake was filled for the first time in 1861. The Sackler Crossing is the first walkway across the Lake and provides a new route through the gardens allowing a close appreciation of lakeside plantings and wildlife.

The four thickly wooded islands of the Lake are important nature conservation areas, undisturbed from regular human activity. The plantings have been planned so that the islands blaze with colour in autumn, reflecting with marked effect in the water. *Nyssa sinensis* turns deep red, while *Nyssa sylvatica* turns red, orange and yellow. On the north side, new swamp cypress (*Taxodium distichum*) have been planted.

Opened in May 2006, the Sackler Crossing is an elegant walkway spanning the Lake; the work of architect John Pawson. Its design fosters clear visual links between the bridge and the natural contours of the gently rounded shoreline, the smooth expanse of the Lake and the powerful verticals of the trees. Set close to the Lake's surface, the Crossing gives an illusion of walking on water and enables you to feel you are a part of nature's waterside scene. Depending on your viewpoint, the spaces between the bronze fins appear and disappear, giving the structure an intriguing ambiguity between solid and transparent, like water itself.

Waterfowl

Many different waterfowl can be seen around the Lake and Palm House Pond, including the Red-crested pochard (pictured above). Also keep an eye out for the tufted duck, widgeon, barnacle goose, bar-headed goose, great crested grebe, shelduck, shoveller, mandarin duck, cormorant, greylag goose, and Egyptian goose.

Waterlily Pond

The Waterlily Pond, created in 1897 from a small gravel pit along Cedar Vista at the edge of the Woodland Glade, sports many flowering plants including water lilies and irises. It is best seen from May to September. While enjoying the Pond keep an eye out for Kew's peacocks and ornamental golden pheasants, which are often found roaming this area.

Riverside views

Syon and Cedar Vistas terminate at a viewpoint across the Thames to Syon Park, and this is a great spot for bird watching. Look for herons feeding along the river margins and roosting in the trees above, and cormorants fishing singly in mid-stream or drying their plumage in the trees. Sparrowhawks and kestrels can be seen and flocks of up to 20 ring-necked parakeets criss-cross the river. In autumn wood pigeons can be seen gorging themselves on acorns in the holm oaks, while jays gather theirs and bury them in the nearby grass.

Wildlife

Dragon and damselflies can often be seen close to clean water, which they need to breed. Around the Lake and Waterlily Pond, blue-tailed and common blue damselflies can be seen from late May. The first dragonflies, broad-bodied chasers, emerge at the same time and are followed by Emperor dragonflies, common, southern, migrant and brown hawkers and lastly common darters. In all, 14 species of dragon and damselflies occur at Kew. The difference between dragonflies and damselflies is best seen when at rest. Dragonflies keep their wings outstretched like small helicopters, while most damselflies fold theirs back along the length of their bodies. Dragonflies are also usually sturdier than slender damselflies.

Wildlife Observatory

More water loving wildlife can be seen at the Wildlife Observatory in the Conservation Area. Dragonflies and damselflies, frogs and newts, and a myriad of water-creatures can be found in this area. A wide range of habitats, plants and wildlife can be seen from the observatory, where the benefit of conserving such habitats and building garden ponds is outlined. A wooden observation hut overlooks the gravel pit in one direction and a dipping pond in the other. Inside there is more information about the habitats and the wildlife they support. The dipping pond has been planted with the aim of attracting aerial insect visitors and as many amphibians as possible. Wooden models portray wildlife visitors may typically encounter: a dragonfly, a frog and a great crested newt.

The Conservation Area

Kew supports an extraordinary diversity of British wildlife, much of which can be found in the Conservation Area around Queen Charlotte's Cottage. Its 15 hectares are actively managed to encourage native flora and fauna.

It is mainly woodland, with many British trees represented including oak, beech, holly and yew. In spring, the woodland is much admired for its carpet of bluebells, wild garlic and snowdrops. There are also some rare native trees, such as the Plymouth pear and the Bristol mountain ash. Very few examples of these are left in the wild, in Devon and the Avon Gorge respectively, and here they are fine examples of Kew's conservation efforts in saving many threatened species from extinction.

Dead trees are left standing or where they fall. Holes and crevices in old, dying and dead trees provide excellent feeding and nesting opportunities for birds and roosting places for Kew's many bats. They are also essential for the myriad fungi to thrive and then eventually break down into valuable nutrients for the soil. More species depend on dying or dead wood than on living trees. Hazel coppicing – cutting the trunks of trees to ground level and letting shoots grow up from the stumps – takes place on a regular cycle. This keeps woodland open, helping understorey plants and wildflowers to gain a foothold. Young hazel shoots grow into long sturdy poles to use for plant supports and woven hurdles in the gardens.

There are meadows and grassy areas, wetlands, ponds and a small gravel pit, all helping to support native butterflies, dragonflies and other insects. Amphibians, too, find a home here. Many of them are in decline due to changes in farming and new building destroying their habitats. Larger mammals, such as foxes and badgers also thrive here.

The Compost Heap

Peat – for years every gardener's choice for mulching, potting and seed compost – is a rapidly dwindling natural resource and is a wildlife habitat. So Kew largely suspended the use of peat in 1989, even though peat substitutes were then in their infancy.

Today, both Kew Gardens and Wakehurst Place have extensive internal composting programmes. Kew royally mixes the 10,000 cubic metres of waste plant material it generates every year with horse manure from some exclusive stables: the Royal Horse Artillery and the Wardens of Windsor Great Park! Watering and turning through a 10–12 week cycle, it produces 2,000 cubic metres of compost for use throughout the gardens. If you want to take a look behind the scenes there is a viewing platform in the Pinetum, close to the Lake.

Bees

During 2009, Kew is placing bee hives just inside the Main Gate area. Bees feed on pollen and nectar and are important pollinators. In the UK about 70 crops are depend on, or benefit from bees, and they also pollinate many flowers. Worldwide, the decline of bees has become a serious issue.

The Stag Beetle Loggery

The intriguing 'Loggery' near Queen Charlotte's Cottage is part of the London Biodiversity Action Plan for Stag Beetles. The Thames Valley is a hotspot for stag beetles and Kew is doing all it can to encourage them.

Wildlife

Birds

Bird experts visiting Kew have reliably listed no fewer than 128 bird species. Trees, grass and scattered shrubs attract the usual birds of London's outer suburbs, such as kestrels and sparrowhawks, woodpeckers, blackcaps, chiffchaffs, nuthatches and treecreepers.

You may find blue, great and coal tits, robins, and friendly introduced pheasants in the dells. There are herons and cormorants on the Thames, while within Kew, great crested grebes nest on the Lake and Palm House Pond, with an annual invasion of wild waterfowl, supplementing the Canada geese and resident ornamental birds.

Butterflies

At least 28 species of butterfly belonging to five families have been recorded at Kew since 1980 – a high number considering the London location, but a reflection of the variety of plant life and habitats within the gardens. Nectar plants are important for the butterflies, but food plants for their caterpillars are just as vital, together with habitat management to allow each species to complete its annual life cycle. The mowing regime of many areas at Kew is designed specifically to provide nectar for the adults and sufficient food to see the caterpillars through to their pupal forms. Kew's butterflies are present at some stage of their life in winter, too.

Fungi – the mysterious partner

Fungi are a bit like icebergs – what you see above is only a small part of the whole. What appears above the ground, tree trunk or rotten vegetation as a mushroom, toadstool or bracket is only the fruiting stage, which is relatively short-lived. The main body of a fungus lives within its source of food, which could be the wood or other plant material, soil, or carrion. Fungi are a separate group, neither plant nor animal, consisting of microscopic threads that form a network that expands out through, and feeds on, organic matter. Fungi are usually the primary decomposers in their habitats, releasing nutrients to the surroundings.

Amazingly, around 80% of all plants grow in mutual association with fungi and many, from orchids to pine trees, will not grow without a partner fungus within their roots – a mycorrhizal association. Scientists at Kew study fungi from around the world, investigating their practical uses as biological controls against pests, with the objective of reducing the use of toxic pesticides.

The Badger Sett

Here you may walk through the giant forked oak branch entrances and find out what a badger's home is really like. Food stores, sleeping chambers and nests are all connected by a warren of tunnels. These tunnels are all at least a metre high, and one, 1.5 metres high, is suitable for wheelchairs.

Art at Kew

Kew holds one of the world's greatest collections of botanical art, totalling over 200,000 items.

In the early days, before photography, precise botanical drawing was vital to make a record of newly discovered species. Out of this came an art form of itself – botanical art. Botanical art continues to have an important role alongside photography in detailing information on new species. Much of the botanical art at Kew has been used in important publications and gone on to grace the walls of many a home from King George's day to the present.

The Shirley Sherwood Gallery of Botanical Art

The Shirley Sherwood Gallery of Botanical Art is the first gallery in the world dedicated solely to botanical art and open year round. This new gallery exhibits valuable and unique works of art from Kew's historic collections and Dr Sherwood's own contemporary collection.

Exhibits from Kew's collection includes work by masters such as Georg D. Ehret, the Bauer brothers and Pierre-Joseph Redouté, together with nineteenth-century artists such as Walter Hood Fitch, one of the world's most prolific botanical artists. Many of the Roxburgh drawings, acquired for Kew by Joseph Hooker from the East India Company in 1858, will be exhibited in this gallery from time to time.

Dr Sherwood started collecting botanical art in 1990, her first purchase was an orchid painted by the artist Pandora Sellars. During her worldwide travels she discovered many other botanical artists and she now has work by over two hundred artists from over thirty countries.

The Shirley Sherwood Gallery is next to the Marianne North Gallery and the two buildings are joined together by the Link Gallery. This Link Gallery displays contemporary botanical works from the Sherwood Collection, on a regularly changing basis. The main gallery runs three exhibitions a year featuring material from the Kew Collection, partner institutions and the Shirley Sherwood Collection, with one exhibition per year based exclusively on the latter.

The building, designed by award-winning architects Walters and Cohen, is intended to have minimal environmental impact, with heating and air conditioning designed to minimise energy consumption and special glass and blinds which automatically react to light.

Featured exhibitions at Kew

Kew has held many major exhibitions over the past decade – from Chapungu in 2000, to Emily Young in 2003, Chihuly in 2005 and Henry Moore in 2007–2008. Kew has hosted the International Garden Photographer of the Year exhibition for several years, and new artistic events are constantly being planned for the Gardens.

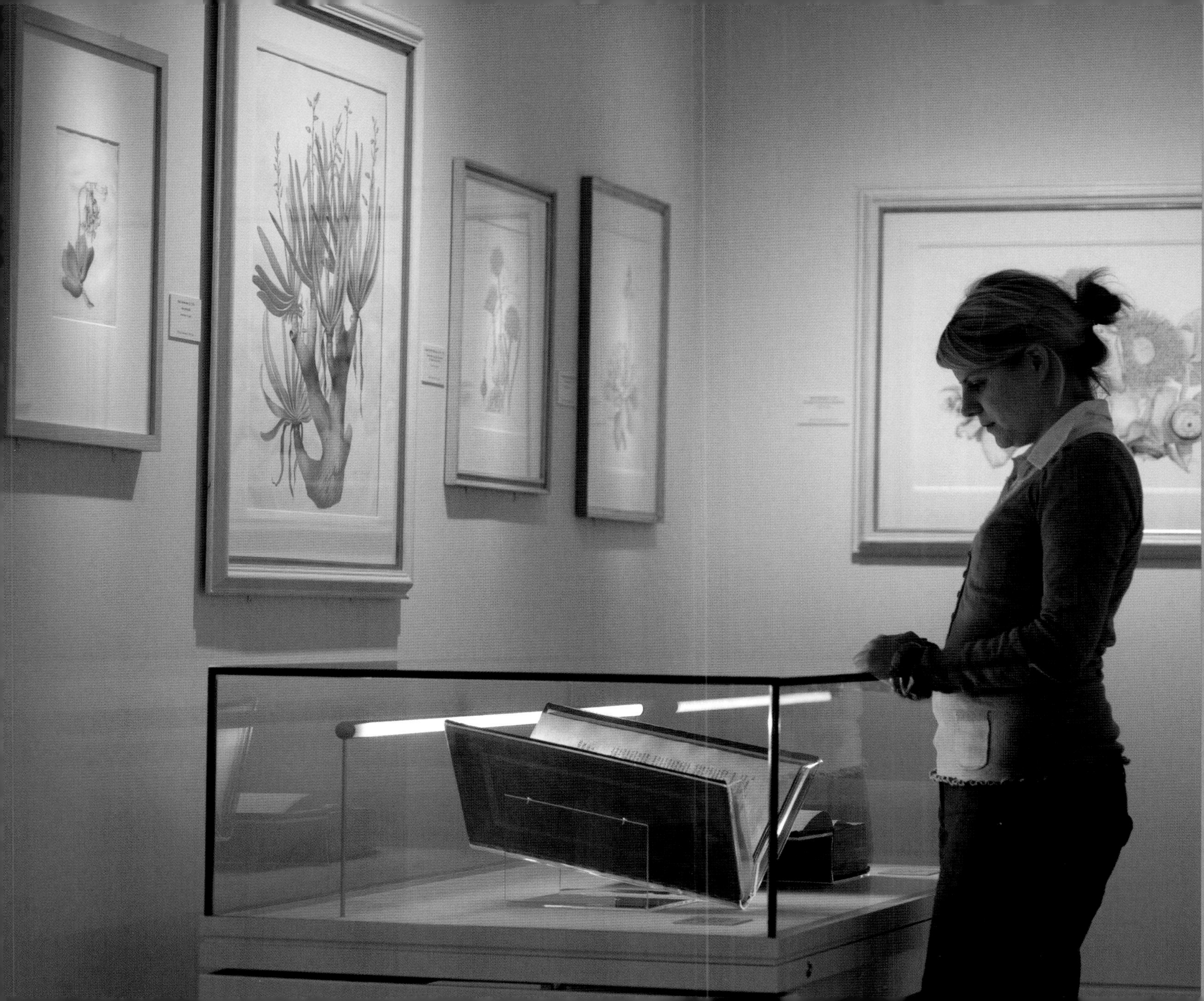

Seed Walk

Throughout 2009, spectacular three metre high willow seed sculptures will be on display between the Main Gate and Nash Conservatory.

The ten sculptures, commissioned from artist Tom Hare, are of various intriguingly-shaped seeds set against the backdrop of three new tropical display beds created for the project. The sculptures have a mild steel framework to give them strength and a firm anchorage. The willow is then woven onto the frame. During the summer, visitors can enjoy seeing some of the sculptures being created on site.

The Kew Gardens Gallery

Located in Cambridge Cottage, Kew Gardens Gallery features botanical art by both past and contemporary artists. This quiet and charming gallery is a favourite for many of Kew's art-loving visitors.

Because of the many facilities in Cambridge Cottage, it may be hired for private functions, such as weddings and other receptions. At these times, the Gallery will be closed to the public.

The Marianne North Gallery

Marianne North was a highly talented Victorian amateur with a great eye for detail. She learnt to enjoy travel with her father, the MP for Hastings, whom she accompanied on his tours of Europe and the Middle East. She was given lessons in flower painting by Valentine Bartholomew, Flower-painter-in-Ordinary to Queen Victoria. On the death of her father, she started her travels, at the age of 40, to search for the world's exotic flora.

Her first solo trip in 1871 took her to Jamaica and North America; her next to Brazil; then Japan in 1875, returning home through Sarawak, Java and Ceylon. India inspired her to paint 200 buildings as well as plants!

Her palette was of bold and assertive colours and her enthusiasm evident, if sometimes rather undisciplined. She sketched rapidly in pen and ink on heavy paper and the oils would come straight from the tube. Somewhat against tradition in Victorian flower painting she liked to paint plants "in their homes" - their natural settings, sometimes with an insect or other small creature.

In 1879, Marianne wrote to Sir Joseph Hooker offering Kew her collection and a gallery to house it. Her only stipulation was a room to use as a studio. Her architect friend James Fergusson designed a gallery which echoed her feelings for India, providing a verandah around the outside while satisfying his own ideas on lighting with large clerestory windows high above the paintings.

Marianne took charge of the hanging, arranging her pictures in geographical sequence over a dado of 246 strips of different timbers. The walls are virtually solid with paintings - there are 832 artworks all told, showing over 900 species of plants - a unique memorial to a quite unique woman.

The Marianne North gallery is currently being refurbished and will re-open in Autumn 2009.

Science and Conservation

Since its inception as a botanic garden Kew has stood for science – both theoretical and practical. Plant hunters scoured the globe for living and preserved specimens and returned them to Kew for study. Many of these surround you in the Gardens.

Over 200 scientists work for Kew – collecting, identifying, investigating, discovering, preserving, conserving. Plants literally give breath to our planet and we need to learn how to use and conserve them. And this ever more urgent work continues.

Global reach

Kew's mission is 'to inspire and deliver science-based plant conservation worldwide, enhancing the quality of life.' This work is of great importance. Plants regulate the Earth's climate, and provide us with the oxygen we need to breathe. We rely on them for food, medicine and fuel. And through providing habitats and food for all living creatures, plants underpin all other forms of biodiversity and are an essential part of healthy ecosystems. In short, all life depends on plants.

The world's vegetation is now greatly diminished, the ice-caps shrinking and deserts expanding. Through its Breathing Planet Programme, Kew is focusing its science and expertise on these critical environmental issues. The Programme focuses not just on expanding our knowledge of the world's plants but on actively providing that expertise to the people who need it most, and working with partners around the world to address environmental issues.

Driving discovery and sharing knowledge

Kew is continuing to make new discoveries about plants and plant diversity, at home and abroad, while ensuring existing knowledge is widely available.

Understanding the whole plant 'mix' – plant diversity – in an ecosystem is of vital importance. Diverse vegetation supports the ecosystem. Plants depend on others – and on animals, fungi and insects – to grow and reproduce. We need to know everything that is 'out there' and why they grow together: without a thorough understanding of the key role of plants, we cannot tackle the environmental challenge.

**Sir Joseph Banks
1743–1820**

Joseph Banks matched an adventurer's heart with an enquiring mind, and with the benefit of a generous inheritance at an early age he found the means to indulge both. He was elected a Fellow of the Royal Society in 1766, aged just 23 and on his first sea voyage. Having whetted his appetite, he then paid for his own team of scientists to join HMS *Endeavour* under Captain James Cook, in 1768.

Banks became a close friend of 'farmer' King George III and from 1773 acted as unofficial 'superintendent' of the Royal Gardens at Kew. He supported many botanical expeditions including the mutinous HMS *Bounty* voyage.

Banks initiated Kew's collection programme, and by encouraging commercial and military enterprises of the British empire to collect plants, ensured a steady supply of new specimens from many far flung parts.

Herbarium, Library, Art and Archives

Every year, thousands of researchers come from around the world to Kew's Herbarium (preserved plant collection) and Library, recently enlarged thanks to a major rebuilding programme. The Herbarium holds over seven million specimens, some going back to the plant hunting voyages of Sir Joseph Banks with Captain Cook.

Nearly half a million of Kew's herbarium specimens have been given priority in our digitisation programme, and to date high-resolution images of nearly 150,000 items have been made available on the web, providing access to these unique resources. The programme also provides database access and online copies of major reference works, including the vast tropical African Floras. Kew is working with London's Natural History Museum and leading US institutions to make core biodiversity literature freely available online – a vital resource, particularly for the developing world.

The Jodrell Laboratory

The first Jodrell Laboratory was built in 1877 to study the internal form, physiology and chemistry of plants. Built in a cottage style it had just four main research rooms and an office. It was put up and equipped for £1,500, donated by a visionary T. J. Phillips Jodrell.

A team of nearly 80 scientists work in the Jodrell, using the latest analytical techniques to study plant families that have economic importance or particular interest, such as Poaceae (grasses), Orchidaceae (orchids), Leguminosae (beans) and Arecaceae (palms). Current research includes: identifying substances to help in the fight against illnesses such as cancer and malaria; DNA barcoding for plants (using genetic markers to distinguish between plant species and allow rapid identification by customs officers and forensic scientists); and plant–animal interactions, especially the host selection behaviour of insects.

The Jodrell also includes the Sustainable Uses Group, which curates a collection of over over 92,000 plant products, implements and artefacts from around the world. Highlights from the Economic Botany Collection are on permanent display at the Plants+People Exhibition in Museum No. 1.

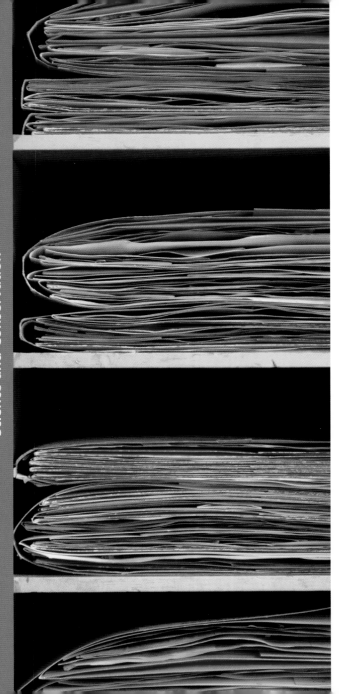

Supporting threatened areas, meeting the challenge of change

Today, Kew is the world's leading authority on global plant diversity and conservation. Our science teams work to identify highly threatened plant and fungal species and threatened regions. Alongside this, practical help is given to global conservation projects. This includes drawing together scientists, ecologists and others to advise on the restoration of badly damaged habitats. Our network of international collaborators consists of over 500 partners in more than 100 countries and includes botanic gardens, NGOs and private sector companies. Support is also directed to areas that have changed permanently, by advising on planting and the most suitable species for the new environmental conditions. This work improves everyday lives around the world now and for the future.

**Sir William Hooker
1785–1865**

Sir William Hooker was appointed as the first 'official' Director at Kew in 1841. He was a dynamic choice, and in him the ailing gardens found a new champion.

Sick plants were revived and trees and shrubs were planted in family groups throughout the site. He constantly increased the collections – and drew wide public interest. Soon out of space he ordered construction of the now world-renowned Palm House in 1848. He initiated the Economic Botany Collection, which led to the building of Museum No. 1, and in 1860 he oversaw the start of the Temperate House construction.

William Hooker, truly the 'father' of the Kew we know today, passed the Directorship to his son, Joseph Dalton Hooker, in 1865.

**Sir Joseph Hooker
1817–1911**

Joseph Hooker, Sir William's son, trained as a doctor in Edinburgh but also loved botany. Between 1839 and 1843, he travelled as assistant surgeon and research botanist on HMS *Erebus*, visiting many places including Madeira and the Cape in South Africa. He identified many previously unknown species of rhododendron during journeys through India and Nepal in 1848–51; some of these can still be seen in Rhododendron Dell today.

By 1865, when he assumed the Directorship at Kew, there were more than 13,000 species in cultivation in the gardens with over 3,000 species of trees and shrubs in the Arboretum. His love of taxonomy led him to introduce the Order Beds, with plants arranged according to the Bentham-Hooker classification. He retired from Kew in 1885.

Madagascan 'suicide palms'

Specialist knowledge is vital, but a new discovery often starts with simple observation. These came together in a recent find in Madagascar: one of the so-called 'suicide palms'.

A family in Madagascar, out for a picnic, were astonished to see what looked like a Christmas tree growing from the top of a large palm. They had seen the palm before and not noticed anything unusual so they made contact with local botanists, and soon there were pictures on a specialist website.

Kew's analysis identified the palm as new to science – in

fact, it is a new genus, *Tahina*, with an affinity to the tribe Chuniophoeniceae found from Arabia to Vietnam. As yet, there is no known lineage connecting it to Madagascar and only a hundred individuals have been identified, making it a conservation priority. The tree died after flowering – hence the suicide headlines. The local community are protecting the trees, and its seeds are a new source of income for them.

Conserving Cameroon's rainforests

Kew's work has led to new national park being created in Cameroon – helping to protect an important conservation area from logging.

Cameroon's forests are part of a rainforest second only in size to the Amazon. In 1995 Kew began working local partners on a full survey of plants in the Bakossi Mountain range, one of the largest areas of cloud forest in Africa. Over 9,000 exceptionally diverse specimens were collected, with about 2,500 species, 89 of which are found nowhere else. Kew has built strong relationships with local people and in 2007 Cameroon's

Prime Minister and Minister for Forests and Environment created the Bakossi National Park. Kew is now working to build a Red Data Book (a compilation of conservation assessments) for the whole country, to help Cameroon manage the survival of its threatened plants.

Madagascar mapped

Madagascar is home to more than 10,000 plant species and, incredibly, 90% of its plants occur nowhere else in the world. Just 18% of Madagascar's native vegetation remains intact – a precarious situation. In 2007, Kew, Missouri Botanical Garden, Madagascar's government and Conservation International published the first vegetation atlas of Madagascar – using maps produced with satellite imagery and other state of the art technologies. This pioneering atlas, the culmination of over 20 years of conservation work, will guide decisions about habitat and forest management in a country with extraordinary species – some of which are already known to have properties of great value to humanity.

Montserrat

Habitat restoration and species reintroduction will become important as the effects of climate change become more marked. But natural disasters are not always due to climate change. Between 1995 and 1997 the Caribbean island of Montserrat was devastated by volcanic eruptions that destroyed the capital Plymouth and much of the island's unique vegetation. With the help of Kew scientists, a new botanic garden has been established in Montserrat. Native vegetation is being grown in the garden for re-introduction into the island's damaged habitats.

Colin Clubbe, leader of Kew's UKOTs (UK Overseas Terrritories) programme, examining a specimen in Montserrat.

Plants are identified, collected, pressed and dried in newspaper before being sent to Kew's Herbarium. In the image below, field identification characteristics are being demonstrated, prior to the plant being pressed. Collecting herbarium specimens is vital work, not least because conservation and sustainable use projects depend upon accurate identification of the plant involved.

Ghana and Kenya – coping with change

Changes in land use, climate change and population growth all threaten the ability of communities to harvest the plants they have traditionally relied on for food, shelter and medicine. Kew's sustainable use team are applying science to help them.

In Ghana, local scientists and communities work with Kew to survey plants used locally to treat malaria. In many areas, people rely on plants as their main form of medicine, with less than 10% having access to modern drugs. Changes in

land use mean that many of the most effective species are now scarce. These projects aim to evaluate the efficacy of the plants and work with communities so they can grow these plants sustainably.

In Kenya, Kew is helping a number of communities to assess the nutritional value and prospects of sustainable horticulture on smallholdings for a range of wild food plants that are now in short supply. Similarly, in South Africa, Kew is helping a local charity supply sustainably grown medicinal plants to communities hit hard by HIV/AIDS.

Safeguarding species through the Millennium Seed Bank partnership

Kew's Millennium Seed Bank partnership is the world's most ambitious plant conservation initiative. It is the largest wild plant seed bank and will eventually hold the seeds of at least half of all known plants. By 2010 seed from 10% of the world's approximate 300,000 wild plant species known to science will be banked, including the rarest, most threatened and most useful species. Already banked are seeds from 96% of UK plant species, and more than 75% of threatened UK plant species. The post-2010 goal is to collect and conserve 25% of the world's wild seed species by 2020.

Seeds to be banked are carefully collected, inspected, germinated to confirm viability and then stored deep in temperature controlled chambers under the Sussex High Weald. Through the MSB exhibition space, the Orange Room, you can watch the scientists at work on this amazing project.

Collecting and conserving wild plant seeds provides an insurance policy against extinction and options for their future use. In the 12 years since its launch, the project has supplied thousands of seeds for research. MSB seeds are put to many uses, including counteracting salt saturation of agricultural lands in Australia; developing drought-resistant forage plants in Pakistan and Egypt; improving and controlling the quality of essential oils in Brazil; and developing new foods in Mexico.

A network of 123 organisations from 54 countries contribute to the seed banking partnership, making it unique in its ability to collect, conserve and study the world's flora. Seeds are also stored in their country of origin, and Kew's support, assisting with facilities, advice and training, is as important as the seed collecting itself.

Botanic gardens worldwide

Central to Kew's ethos is the belief that botanic gardens are places to inform and inspire people. Visitors, be they researchers, plant lovers, plant scientists, artists, journalists, schoolchildren, adults or senior citizens, can be amazed and educated by the power – and vulnerability – of our planet's plants. Botanic gardens across the world are ideally placed to deliver enjoyable, inspiring experiences, helping to create a well-informed general public who will care for the global environment.

Wakehurst Place

Kew's country estate is set in the beautiful High Weald of Sussex. The elegant Elizabethan mansion is surrounded by formal gardens, woodland walks, nature reserves and is the home of Kew's Millennium Seed Bank.

The climate at Wakehurst Place is milder than Kew's. Higher rainfall and more moisture-retentive soils allow many important groups of plants, which struggle at Kew, to flourish including conifers, rhododendrons and an outstanding collection of temperate flora from eastern Asia, South America, Australia and New Zealand.

Gerald Loder (1861–1936) began the botanical collections which made Wakehurst Place famous. On Loder's death, Wakehurst Place was bought by Sir Henry Price and under his care the estate matured. Sir Henry left Wakehurst Place to the nation in 1963, which gave Kew the opportunity to lease it from the National Trust in 1965.

The Great Storm of 1987 effectively removed most of Loder's collections. Over 20,000 trees were lost on that October evening. Since then, Kew has redesigned the landscape and collections to create an 80 minute walk around the temperate woodlands of the world. The walk takes you through New England's

stunning autumn colour, California's majestic giant redwoods, Australia's magnificent eucalypts, Patagonia's southern beeches and the Himalaya's beautiful birches and rhododendrons.

Wakehurst is a garden for all seasons with its Winter Garden, Spring Border and summer displays in the Walled Garden and the extensive Water Gardens. The tree and shrub collections are ablaze with colour in autumn. Additional year-round interest is displayed in the Tony Schilling Asian Heath Garden, the Southern Hemisphere Garden, and the four National Collections – birches, southern beeches, skimmias and hypericums.

Loder Valley Nature Reserve

The Loder Valley Nature Reserve encompasses three major types of habitat; woodland, meadowland and wetland. It conserves different habitats complete with their own birds, mammals and invertebrates through traditional countryside management. Coppicing, for example, not only produces wood for Bar-B-Kew charcoal and country products such as hurdles, but also restores a habitat for the native dormouse.

The Francis Rose Reserve

The Francis Rose Reserve is probably the first nature reserve in Europe to be dedicated to the conservation of cryptogams (mosses, liverworts, lichens and filmy ferns). Francis Rose was a renowned botanist who pioneered the study of these plants and the sandrock outcrops of the Sussex High Weald where they grow. A new hide, opened in late 2006, allows greater enjoyment of the wildlife here.

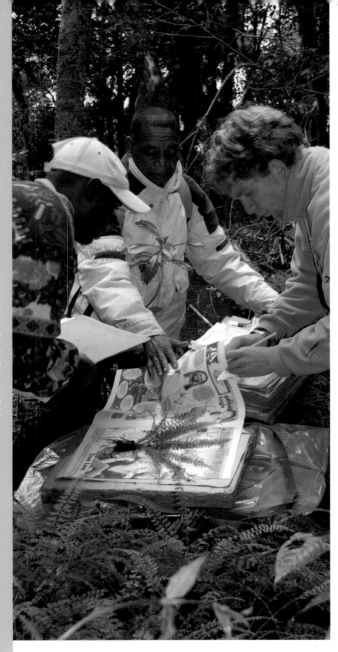

How Kew is funded

The Royal Botanic Gardens, Kew, is a registered charity established in April 1984 under the terms of the National Heritage Act. Governed by a Board of Trustees, Kew is partially supported by grant-in-aid from Defra (Department of Environment, Food, and Rural Affairs).

As environmental challenges become ever more acute, the demands on Kew's work intensify. Although Kew receives approximately half of its income from central government, it is the investment of voluntary income from visionary organisations and individuals that enables Kew to respond to these growing challenges. Kew Foundation is instrumental in securing this investment.

Established in March 1990, the mission of the Foundation & Friends of the Royal Botanic Gardens, Kew (the Foundation) is "to develop and sustain a strong network of support in order to raise awareness and funds for Kew's cause worldwide".

Voluntary income enables Kew to work ahead of government, embarking on ambitious and challenging plant science and conservation projects in the UK and around the world. From building and running the Millennium Seed Bank to training botanists in the heart of Africa, from digitising Kew's vast collections to developing education programmes to reach every child in the UK – private donations are integral to Kew fulfilling its vision. Without this support, Kew could only deliver a fraction of its work and would have to significantly scale back activities at a time when its resources and expertise are needed more than ever.

Kew runs active and varied Volunteer and Community Outreach programmes. The Community Outreach programme fosters links with our neighbouring communities and engages people with the urgent need for conservation and sustainable use of natural resources. Our volunteers play a vital role in supporting Kew and are an integral part of the organisation. About 500 volunteers work across many areas: from helping young children explore Climbers and Creepers, to working as school explainers, engaging and involving children with exhibits around the gardens, and assisting with the Discovery Programme for elderly and disabled people. Volunteers also run the information desk at Victoria Plaza and undertake basic horticultural maintenance in the Gardens. If you are interested in future volunteer opportunities please visit *www.kew.org/aboutus/ volunteers* for further information.

Kew membership

With over 80,000 members worldwide, the Royal Botanic Gardens, Kew has a strong and growing network of supporters who recognise the vital role Kew has to play in protecting and preserving our most fragile asset: our diverse and irreplaceable plant kingdom. Kew recognises the support of its members with free entry to the magnificent Gardens at Kew and Wakehurst Place plus exclusive membership benefits including free guest passes for family members, access to the latest information about activities on-site and much more. Becoming a member of Kew is excellent value for money and is suited for individuals and families alike. Kew membership also makes for a unique present, appropriate for any special occasion.

Joining is easy and convenient: you can buy your membership at the gates, at the information desk at Victoria Plaza or over the phone or the internet. Call the membership office on 020 8332 3200 (9am to 5pm) or visit www.kew.org for full details.

Kid's Kew

Nearly 200,000 children visit Kew each year, either as part of a school group, or on a great-value family day out (children under 17 have free entry to the gardens). 'Education' perhaps sounds stuffy and formal but here at Kew it is anything but. Our mission in education is to make the plant kingdom interesting and fun and its importance understood. If impressed when young, children may become conservationists for life.

Certain garden highlights are particularly appealing to younger visitors: from the Xstrata Treetop Walkway (p.54) to the Badger Sett (p.65), the Stag Beetle Loggery (p.65), and Evolution House (p.23), to name just a few. We also have trails designed specifically for children.

If your children are aged between 3 and 9, be sure to visit Climbers and Creepers, Britain's first interactive botanical play zone. Though you might want to explore other areas of the garden first, as kids tend to not want to leave! Children find botany engaging in 'Climbers and Creepers'. Here, they can climb inside a plant to learn about pollination, see the dangers insects face when they are 'eaten' by a giant pitcher plant, crawl through a bramble tunnel and have a lot more fun while learning about plants and their relationships with animals and people. And they can carry on the fun at home by visiting *www.kew.org/climbersandcreepers*, which features interactive games and activities and all the Climbers and Creepers characters.

Please note: while properly-trained and security-cleared Kew staff and volunteers work in the zone, parents and minders remain fully responsible for their children, and are advised to keep an eye on them at all times.

Kids' Kew is an action-packed, interactive guide to the gardens, aimed at 7–11 years olds, featuring exciting places to visit and things to do at all times of the year. It is available to buy from any ticket kiosk or shop at Kew.

Education: the ladder of learning

The Kew learning team runs an exciting programme of events and activities. Find out how to book activities and workshops by checking out the website www.kew.org/education/schools. School visits to the Gardens provide children with a high-quality, hands-on experience. On a typical visit children get the opportunity to investigate the way plants adapt to different environments, learn about the uses of plants and explore some of the problems faced by the natural world.

Kew has developed, and is expanding, its range of curriculum resources:

The Great Plant Hunt is the UK's biggest school science project. As part of a programme commissioned and funded by the Wellcome Trust, every state primary school in the UK has been sent a Darwin Treasure Chest jam-packed with outstanding free resources. Activities include exploring habitats, collecting seeds and growing plants. Children are helping with real scientific experiments, and information and seeds gathered are being sent to researchers at Kew's Millennium Seed Bank. Visit www.greatplanthunt.org for more information.

Tree resources produced by Kew and the Woodland Trust support school visits to Kew and help plan activities centred on trees, either at Kew or in a local habitat. Resources cover all key stages and can be searched by key stage or by subject. See www.kew.org/education/index.html

To help children make the most of their visit to Kew, we have developed a number of garden trails, tailored to specific key stages; these can be downloaded from our website. Teachers packs to help plan pre- and post-visit work are also available to download or purchase.

Higher Education

Higher education and training includes the Kew Diploma in Horticulture and the international diploma courses in Botanic Garden Management, Herbarium Techniques, Plant Conservation Strategies and Botanic Garden Education. Kew is a partner in various MSc programmes and runs a PhD programme.

Public Education

Our wide-ranging public education programme includes short courses and study days, tours, lectures, trails, events, exhibitions, and includes the many signs and notices around the Gardens giving full information about the plants you can see. We offer a varied range of short-courses: covering subjects such as photography and botanical illustration, through to growing orchids and even bushcraft skills! For a full listing of current courses check our website or email adulted@kew.org.

Seasonal Walks

Kew is a seasonal pleasure, featuring an array of changing delights as the year progresses: from bluebells in spring to the sweet bloom of summer, from glorious autumn colour to the icy beauty of winter. See the Gardens at their seasonal best by following one of our themed walks. Each walk takes between 2–3 hours at a gentle pace, depending on how long you spend in the glasshouses and on detours.

Spring Walk

1 Go through Victoria Plaza, past the Campanile, and immediately turn left towards the Palm House. Turn right to walk the length of the Palm House, with the spring bedding displays on your left. There may be some waterfowl displaying on the Pond hoping to attract a mate.

Snowdrops (*Galanthus nivalis*) signal the end of winter and, with the right weather, can continue blooming into March.

2 Turn right, away from the Palm House, go straight on at the roundabout and the Woodland Garden is ahead, bursting into life. After sampling its delights you can either go through the Princess of Wales Conservatory to the north end, where there's an attractive seasonal display, or walk through the Rock Garden, where there's not only a superb collection of snowdrops and other spring flowers but also the award-winning Davies Alpine House.

3 From the Princess of Wales Conservatory north exit, go straight ahead to where you see a huge stone pine tree. Follow the brick path on your left into the Secluded Garden. Walk through the garden to the other side where, a little to the left, there's a wonderful display of wisteria on an old pergola. Next to this stands the maidenhair tree (*Ginkgo biloba*) one of Kew's 'Old Lions' (see page 58 for more on Kew's heritage trees).

The beautiful *Davidia involucrata* is best seen in May, when it is in full flower.

4 As you leave the Secluded Garden head right towards the Main Gate. Just past the path to the Orangery there's a planting of magnolias and a spectacular *Cornus* 'Ormonde', which shouldn't be missed.

(5) Go to the Orangery – maybe stopping there for a little light refreshment. Look down the Broad Walk; in February and March you'll see a mass of daffodils on either side – Wordsworth's 'golden host' is a perfect description.

(6) With the Broad Walk on your left, head up the main path to an intersection by the pretty Lilac Collection, at its best in May. Take the path forking left towards a large cedar tree, and stroll along past wild daffodils, crocuses, snake's head fritillaries and a magnificent display of magnolias.

(7) Turn right across the grass, following the Azalea Garden signs. The azaleas are an eye-opener in April and May.

(8) Walk over the grass into the Azalea Garden. After reaching the circle of beds, turn right to join the path heading down towards the river, the Bamboo Garden and Rhododendron Dell.

(9) There's a choice here – either take the path left to the Bamboo Garden, with the fascinating traditional Japanese Minka, or the one taking you through Rhododendron Dell. On the other side of these gardens take the path heading left to Syon Vista. Stroll down the vista towards the Palm House and then turn right onto the elegant Sackler Crossing over the Lake. On the other side of the Lake, join the main path, and, if you have time, turn left and follow the signs to the Rhizotron and Xstrata Treetop Walkway to gain a stunning aerial view of the gardens. Return to this point and head towards the Badger Sett.

Enjoy reflections of lakeside plantings and the Crossing itself while walking over the Sackler Crossing.

(10) Turn left at the signpost to Queen Charlotte's Cottage. The Badger Sett is on the left. Continue through the woods and look out for the Wildlife Observatory on the right. Continue past the Cottage with its bluebells to the dramatic Stag Beetle Loggery. Continue on this path until you reach a crossing of the paths.

(11) If you have time for a diversion, turn left to see some magnificent giant redwoods – otherwise carry straight on through the bright lime-green larches towards the Temperate House.

(12) Fork left towards the Temperate House where the crab apple trees are in flower. Either detour through the house, or continue round the outside, but head left towards King William's Temple and the mass of blossom along Cherry Walk, with beautiful blue scillas planted under the trees.

(13) At the crossroads after the temple, turn right (signpost Victoria Gate) and pass the Temple of Bellona, with a massed carpet of crocuses all around.

(14) Carry on down this path and, as you reach the T-junction by the Kew Road wall, look right at the camellias. Just beyond there are the Shirley Sherwood and Marianne North Galleries with their superb paintings. Go and admire them, or turn left back to the Victoria Plaza.

Summer Walk

1 Go through Victoria Plaza, pass the Campanile, turn right at the Temple of Arethusa and make your way round the Pond to the Plants+People Exhibition. Turn sharp right at the end of the building and with the Temple of Aeolus on its hillock to the left; follow the path to the peony beds.

2 Past the peonies are the Order Beds, where related plants are grouped together in a colourful parade. The students' vegetable gardens (p.46) are also here.

3 Leave the Order Beds by the centre left exit and you are faced with the colourful Rock Garden. Head right, through the centre of the Rock Garden to the award-winning Davies Alpine House.

4 The Grass Garden has over 550 species on view, from fine lawn through cereals to bamboos. Next to it is the Duke's Garden with the sweetly-scented Lavender Trail and the Gravel Garden, full of ideas for plants that need very little water.

5 Turn right out of the Duke's Garden and on the left there's a path to the Princess of Wales Conservatory. If you have time, enjoy the seasonal display where you enter and a magnificent giant waterlily which should be in full flower in the centre of the house.

Princess of Wales Conservatory

6 Continue along the path to the huge stone pine tree. Opposite is a sign leading into the Secluded Garden, with plants to stimulate all the senses, a conservatory and a bamboo tunnel.

7 Go through the Secluded Garden and turn right to the Main Gate, where you then turn left along a path lined with magnificent Indian horse chestnut trees, with candles of flowers proud on their branches.

Aesculus indica

8 Pass the rear of the Orangery, turn right and pass the Kew Palace Welcome Centre. Kew Palace is well worth a visit. The elegant Queen's Garden, with its 17th century inspried parterre, a nosegay garden and a splendid laburnum arch, is behind the Palace.

9 Leaving Kew Palace, cut diagonally to the right across the grass to White Peaks, where you might stop for a bite to eat and drink or to visit 'Climbers and Creepers', a perfect break for children. When ready, head straight on to the Lilac Collection and continue on to join the Broad Walk.

10 Turn right along the Broad Walk and right again at the roundabout for the Palm House. Turn right into the Waterlily House to see a giant waterlily, loofahs, papyrus and the sacred lotus.

11 Inside and either side outside the Palm House are joys. There's summer bedding on the Pond side and a vast display of roses on the other. Enjoy any or all of them, then take the path away from the roses in the direction of the Temperate House.

12 Soon, there's a crossroads in front of King William's Temple, with its specially planted collection of highly-scented Mediterranean style shrubs, herbs and other plants.

13 From the crossroads turn right until you reach where five paths meet. Continue virtually straight ahead and then fork left to the Rhizotron and Xstrata Treetop Walkway, where you can explore deep in the ground and then high up in the sky.

14 Return to the path and cut across the grass to the path by the lakeside. Just over halfway along, take the path on the right to the elegant Sackler Crossing. Cross the Lake and turn left onto Syon Vista. From here it is only a short detour to explore the Bamboo Garden and Japanese Minka. Otherwise, head up Syon Vista towards the river and then around the top of the Lake.

The Japanese Minka in the Bamboo Garden

15 From the head of the Lake, go over a grass hummock, across the path and down Cedar Vista towards the Pagoda. On the right is the Waterlily Pond and just past it on the left, the seclusion of the beautifully planted Woodland Glade.

16 Carry on down Cedar Vista over several paths to the tranquillity of the Japanese Gateway and Landscape on the right and then, just ahead, the Pagoda reaches up to its full ornate height.

17 Turn left and walk the length of the Pagoda Vista – don't forget to look behind you to appreciate the view. At the Temperate House turn right onto a path towards the Shirley Sherwood and Marianne North Galleries. Why not stop and have a look around them? Then continue on to the T-junction and turn left to return to Victoria Plaza.

Autumn Walk

1. Exit Victoria Plaza near the ticket booths and turn left onto a path parallel to Kew Road, where the smoke bush is changing colour by the Temple of Bellona.

2. Continue along the path, turn right and walk through a blaze of autumn colour among the maples towards the Temperate House.

3. Before you reach the Temperate House, turn left on to the grass of Pagoda Vista, and stroll through more glorious autumn colour.

There is no better place to view the Arboretum in all its autumnal glory than from the top of the Xstrata Treetop Walkway.

4. Turn right at the first path (away from the Pavilion Restaurant), and go past the end of the Temperate House to a T-junction. Turn right onto Holly Walk and walk along to the left turn opposite Evolution House. Head up this turning to the Rhizotron and Xstrata Treetop Walkway. From the top of the walkway the autumn colours take on a new dimension! After exiting the walkway, take the path towards the broad spread of Cedar Vista.

5. Walk up Cedar Vista, through the glorious autumn colour, past the Waterlily Pond. Don't forget to look back at the view.

6 At the end of Cedar Vista you meet a path, where you turn right. When you reach the Sackler Crossing, stroll over to enjoy the autumnal reflections in the Lake. Cross Syon Vista, and head through the trees towards the Bamboo Garden and Rhododendron Dell.

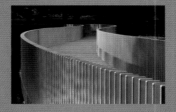

Cyclamen

7 In the centre of the Bamboo Garden is the intriguing Japanese Minka. Take the central path of the Rhododendron Dell and then carry straight on to another path, where you turn right to reach the Broad Walk. There, a left turn to the Orangery would bring you some light refreshments. If you have a little extra time (about 20–30 minutes), you could see the beautiful autumn-flowering cyclamen in the Queen's Garden behind Kew Palace.

Rich autumnal hues can be seen in Kew's Arboretumn throughout September and October.

10 Now head right towards Museum No. 1 and the Plants+People Exhibition, which is really worth seeing, so try to make the time. To end your walk, continue round the Pond opposite the Palm House and turn left at the Temple of Arethusa to the Victoria Plaza.

8 Halfway down the Broad Walk turn right, towards the Princess of Wales Conservatory. The Japanese pagoda tree is on the right. Turn right at the historic urn and head to the Princess of Wales Conservatory with its seasonal display in the north end. Beyond the Conservatory you can visit the Davies Alpine House on one side and the Grass Garden on the other.

9 Head through the Davies Alpine House and take the path through the Rock Garden. At the end of the main path turn left to the Order Beds, which show just how attractive serried ranks of seed heads can look.

Berberis berries add to the warm autumn colours, and are also an attractive food for birds.

The red oak (*Quercus rubra*) is so named due to its fantastic autumn leaf colour. Although often thought to be an English tree, this oak is native to eastern North America.

93

Winter Walk

① On leaving Victoria Plaza you can enjoy several trees with beautiful winter bark. Go past the Temple of Arethusa and turn right at the Palm House Pond along a glorious row of sweetly-scented winter-flowering viburnum to the Plants+People Exhibition.

Viburnum x bodnatense

② Turn right at the end of the building, and pass the Temple of Aeolus, where you will find early daffodils on the mound if spring has come early, flowering hellebores, daphne and snowdrops in the Woodland Garden. Follow the path round past the Order Beds and, after the wall, turn right into the Rock Garden, taking its centre path.

③ Take this path to the award-winning Davies Alpine House and on to the Grass Garden and its sculptural seed heads. Go through the grasses to the path by the Duke's Garden where you'll find wintersweet. Turn left and go up the path to the stone pine and turn left again to the Princess of Wales Conservatory where there is an interesting seasonal display.

④ On turning right at the conservatory there's a paper-bark maple on the corner, with stunning dark brown bark peeling off to reveal layers of vibrant orange beneath.

⑤ Go up to the T-junction and turn left past the Japanese pagoda tree. Keep left for the Ice House and Winter Garden with its fragrant flowers. Continuing round, turn right and head for the roundabout and then on to the Palm House. The Pond outside usually has a number of ornamental wildfowl preening themselves.

⑥ Go through the lush warmth of the Palm House, taking in the Marine Display in the basement for a fascinating glimpse of four marine environments; and out past Kew's oldest pot plant, through the far (south) door and, on leaving, turn right.

⑦ Take the first path left towards King William's Temple, around which there are strawberry trees on the Lake side and a collection of witch hazels (in bloom from December to February) on the other, as well as many other intriguing plants.

⑧ Return to the path and follow the signs to the Rhizotron and Xstrata Treetop Walkway. It may be chilly but the views from the top will take your breath away. When you are finished return to the path and head towards the Lake.

9 Follow the signpost to the Lake. Take the right fork to the Lake, where you may see some interesting waterfowl. Halfway down lies the elegant Sackler Crossing, worth a detour for an excellent view of the Lake. Just beyond, on the left, is the path to the Compost Heap viewing platform.

10 At the head of the Lake, past Cedar Vista, take the path that forks left and follow the signs to Lion Gate and the Pagoda, which take you through the majestic giant redwoods.

11 Take the second left to pass the Japanese Gateway on your right. Keep straight ahead along Holly Walk until you fork right and go in through the middle doors of the Temperate House.

12 Inside the Temperate House, you'll find the spectacular bird of paradise flowers in full bloom, along with *Banksia* and sub-tropical rhododendrons.

13 Leave by the middle doors on the Kew Road side and walk straight ahead towards the Marianne North Gallery. Get warmed up by visiting the galleries: the Shirley Sherwood Gallery for beautiful delicate botanical artworks from the Sherwood and Kew collections; the Marianne North Gallery for its collection of bold and evocative flower paintings.

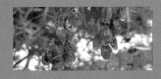

14 Keep to the path (signposted Victoria Gate) past Berberis Dell, which will have some eye-catching winter berries on show. Victoria Plaza is a few yards further on.

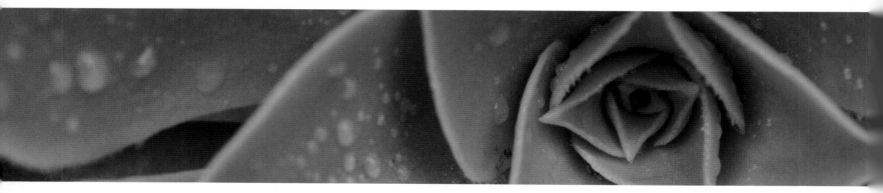

This 4th edition first published in 2009 by the Royal Botanic Gardens, Kew, Richmond, Surrey TW9 3AB, UK
Copyright © The Board of Trustees of the Royal Botanic Gardens, Kew 2009

Editions 1–3
Concept by Paul Cloutman and Michael O'Callaghan. Original text by Paul Cloutman, London SW11, UK
Revised 2007 by Christina Harrison and Paul Cloutman
Designed by Michael O'Callaghan at CDA Design Worthing

4th edition
New material and text revision by Clive Langmead
Edited by Michelle Payne
Designed by Leona Sharkey

Mixed Sources
Product group from well-managed forests, and other controlled sources
www.fsc.org Cert no. CQ-COC-000015
© 1996 Forest Stewardship Council

By buying products with an FSC label you are supporting the growth of responsible forest management worldwide.

Printed and bound in Italy by Graphicom Srl., an FSC (Forest Stewardship Council) and ISO 14001 Environmental Management Certification accredited company.
ISBN 978-1-84246-414-4

All photography by Andrew McRobb except where specified below.
We would like to thank the following for providing photographs and for permission to reproduce copyright material: Cover: front and back flap, Andrew McRobb; front flap, Paul Little; back cover, Jeff Eden. Inside: p11, *Piper nigrum*, Lloyd Kirton; p13 algae images, Heather Angel/Natural Visions; p19 *Doryanthes palmeri* and *Encephalartos woodii*, James Morley; p20, Temperate House, James Morley, p24 Hina Joshi; p 29 Christina Harrison; p37 William Chambers, National Portrait Gallery, London; p41 Decimus Burton, Hastings Museum and Art Gallery; p49 artist's impression of Rose Garden, David Shipp; p65, Bees, Paul Little, p70, seed sculptures, Tom Hare; p75 Joseph Banks, Lincolnshire County Council: The Collection; p79 *Tahina spectabilis* image, John Dransfield; p82 school explainer, James Morley; p94 *Viburnum* x *bodnatense*, Lloyd Kirton; p95 Shirley Sherwood Gallery, Paul Little, *Berberis* berries, Lloyd Kirton.

Profits from the sale of this book will help fund the Royal Botanic Gardens, Kew's global plant science and conservation work.

www.kew.org